Endangered Species
OPPOSING VIEWPOINTS®

Other Books of Related Interest

Opposing Viewpoints Series

    Africa
    America Beyond 2001
    American Values
    Animal Rights
    Biomedical Ethics
    The Environmental Crisis
    Genetic Engineering
    Global Resources
    Population
    The Third World
    Trade
    21st Century Earth
    Water

Current Controversies Series

    Energy Alternatives
    Ethics
    Hunger
    Pollution
    Reproductive Technologies

At Issue Series

    Environmental Justice

Endangered Species

OPPOSING VIEWPOINTS®

David Bender & Bruno Leone, *Series Editors*

Brenda Stalcup, *Book Editor*

Greenhaven Press, Inc., San Diego, CA

No part of this book may be reproduced or used in any form or by any means, electrical, mechanical, or otherwise, including, but not limited to, photocopy, recording, or any information storage and retrieval system, without prior written permission from the publisher.

Cover photo: Dave Plummer, Photodisk

Greenhaven Press, Inc.
PO Box 289009
San Diego, CA 92198-9009

Library of Congress Cataloging-in-Publication Data

Endangered species : opposing viewpoints / Brenda Stalcup,
  book editor.
    p.   cm. — (Opposing viewpoints series)
  Includes bibliographical references (p.   ) and index.
  ISBN 1-56510-365-3 (lib. ed. : alk. paper) —
ISBN 1-56510-364-5 (pbk. : alk. paper)
  1. Endangered species. 2. Biological diversity conservation.
3. Endangered species—United States. 4. Biological diversity
conservation—United States. I. Stalcup, Brenda. II. Series:
Opposing viewpoints series (Unnumbered)
QH75.E66   1996
574.5'29—dc20   95-239
                                                                            CIP

Copyright ©1996 by Greenhaven Press, Inc.
Printed in the U.S.A.

Every effort has been made to trace the owners of copyrighted material.

"Congress shall make no law . . . abridging the freedom of speech, or of the press."

First Amendment to the U.S. Constitution

The basic foundation of our democracy is the First Amendment guarantee of freedom of expression. The Opposing Viewpoints Series is dedicated to the concept of this basic freedom and the idea that it is more important to practice it than to enshrine it.

Contents

	Page
Why Consider Opposing Viewpoints?	9
Introduction	12

Chapter 1: Is Extinction a Serious Problem?

Chapter Preface	16
1. Extinction Is a Serious Problem *Edward O. Wilson*	17
2. Extinction Is Not a Serious Problem *Julian L. Simon & Aaron Wildavsky*	24
3. Deforestation Decreases Biodiversity *John C. Ryan*	33
4. Deforestation's Effects on Biodiversity Are Exaggerated *Shawn Carlson*	41
5. The Extinction of Livestock Breeds Is a Serious Problem *Lisa Drew*	48
6. The Extinction of Wild Plants Threatens Agriculture *Al Gore*	56
Periodical Bibliography	63

Chapter 2: Can Endangered Species Be Preserved?

Chapter Preface	65
1. The Endangered Species Act Is a Failure *Robert E. Gordon*	66
2. The Endangered Species Act Is Effective *Michael J. Bean*	73
3. Preserving Ecosystems Will Save Endangered Species *Douglass Lea*	80
4. Preserving Ecosystems Will Violate Property Rights *Ike C. Sugg*	85
5. Wildlife Reintroduction Programs Help Endangered Species *Colin Tudge*	91
6. Wildlife Reintroduction Programs Harm Endangered Species *Fiona Sunquist*	97

7. All Endangered Species Should Be Preserved 104
   Roger L. DiSilvestro
8. Some Endangered Species Cannot Be Saved 110
   Suzanne Winckler
Periodical Bibliography 117

Chapter 3: Should Endangered Species Take Priority over Jobs, Development, and Property Rights?

Chapter Preface 119

1. Protection of Endangered Species Harms Private Property Owners 120
   Robert J. Smith
2. Private Property Regulation Is Necessary to Save Endangered Species 125
   Bruce Babbitt
3. Saving the Northern Spotted Owl Costs Loggers Their Jobs 130
   Randy Fitzgerald
4. The Northern Spotted Owl Is Not Responsible for Loggers' Unemployment 136
   Alexander Cockburn & Timothy Egan
5. Wolf Reintroduction Threatens Ranchers' Livelihoods 143
   Alston Chase & Human Events
6. Wolf Reintroduction Does Not Threaten Ranchers' Livelihoods 149
   Renée Askins
7. Wetland Regulations Are Unfair to Property Owners 154
   William F. Jasper
8. Wetland Regulations Are Fair to Property Owners 161
   Douglas A. Thompson & Thomas G. Yocom
Periodical Bibliography 168

Chapter 4: What Should Be U.S. Policy on Endangered Species in Other Countries?

Chapter Preface 170

1. The United States Should Condemn Commercial Whaling 171
   Patricia A. Forkan
2. The United States Should Accept Commercial Whaling 176
   Mari Skåre

3. The International Ivory Ban Should Be Maintained 183
   The Animals' Voice & Andre Carothers
4. The International Ivory Ban Should Be Lifted 188
   Richard C. Morais
5. Buying Rain Forest Products Preserves Biodiversity 193
   Diane Jukofsky
6. Buying Rain Forest Products Does Not Preserve Biodiversity 200
   Stephen Corry

Periodical Bibliography 205

Chapter 5: Are Humans an Endangered Species?

Chapter Preface 207

1. Indigenous Peoples Are Endangered 208
   Art Davidson
2. Indigenous Peoples Are Adapting to the Modern World 214
   Alan Thein Durning
3. Humans Are Innately Connected to Nature 222
   Edward O. Wilson
4. Humans Are Not Innately Connected to Nature 227
   Jared Diamond
5. Humans Are Endangering Themselves 234
   Robert M. McClung
6. Humans Are Not Endangering Themselves 241
   Thomas Palmer

Periodical Bibliography 249

Glossary 250
For Further Discussion 251
Organizations to Contact 253
Bibliography of Books 257
Index 259

Why Consider Opposing Viewpoints?

"The only way in which a human being can make some approach to knowing the whole of a subject is by hearing what can be said about it by persons of every variety of opinion and studying all modes in which it can be looked at by every character of mind. No wise man ever acquired his wisdom in any mode but this."

John Stuart Mill

In our media-intensive culture it is not difficult to find differing opinions. Thousands of newspapers and magazines and dozens of radio and television talk shows resound with differing points of view. The difficulty lies in deciding which opinion to agree with and which "experts" seem the most credible. The more inundated we become with differing opinions and claims, the more essential it is to hone critical reading and thinking skills to evaluate these ideas. Opposing Viewpoints books address this problem directly by presenting stimulating debates that can be used to enhance and teach these skills. The varied opinions contained in each book examine many different aspects of a single issue. While examining these conveniently edited opposing views, readers can develop critical thinking skills such as the ability to compare and contrast authors' credibility, facts, argumentation styles, use of persuasive techniques, and other stylistic tools. In short, the Opposing Viewpoints Series is an ideal way to attain the higher-level thinking and reading skills so essential in a culture of diverse and contradictory opinions.

In addition to providing a tool for critical thinking, Opposing Viewpoints books challenge readers to question their own strongly held opinions and assumptions. Most people form their opinions on the basis of upbringing, peer pressure, and personal, cultural, or professional bias. By reading carefully balanced opposing views, readers must directly confront new ideas as well as the opinions of those with whom they disagree. This is not to simplistically argue that everyone who reads opposing views will—or should—change his or her opinion. Instead, the series enhances readers' depth of understanding of their own views by encouraging confrontation with opposing ideas. Careful examination of others' views can lead to the readers' understanding of the logical inconsistencies in their own opinions, perspective on why they hold an opinion, and the consideration of the possibility that their opinion requires further evaluation.

Evaluating Other Opinions

To ensure that this type of examination occurs, Opposing Viewpoints books present all types of opinions. Prominent spokespeople on different sides of each issue as well as well-known professionals from many disciplines challenge the reader. An additional goal of the series is to provide a forum for other, less known, or even unpopular viewpoints. The opinion of an ordinary person who has had to make the decision to cut off life support from a terminally ill relative, for example, may be just as valuable and provide just as much insight as a medical ethicist's professional opinion. The editors have two additional purposes in including these less known views. One, the editors encourage readers to respect others' opinions—even when not enhanced by professional credibility. It is only by reading or listening to and objectively evaluating others' ideas that one can determine whether they are worthy of consideration. Two, the inclusion of such viewpoints encourages the important critical thinking skill of objectively evaluating an author's credentials and bias. This evaluation will illuminate an author's reasons for taking a particular stance on an issue and will aid in readers' evaluation of the author's ideas.

As series editors of the Opposing Viewpoints Series, it is our hope that these books will give readers a deeper understanding of the issues debated and an appreciation of the complexity of even seemingly simple issues when good and honest people disagree. This awareness is particularly important in a democratic society such as ours in which people enter into public debate to determine the common good. Those with whom one disagrees should not be regarded as enemies but rather as people whose views deserve careful examination and may shed light on one's own.

Thomas Jefferson once said that "difference of opinion leads to inquiry, and inquiry to truth." Jefferson, a broadly educated man, argued that "if a nation expects to be ignorant and free . . . it expects what never was and never will be." As individuals and as a nation, it is imperative that we consider the opinions of others and examine them with skill and discernment. The Opposing Viewpoints Series is intended to help readers achieve this goal.

David L. Bender & Bruno Leone,
Series Editors

Introduction

"All other environmental problems pale beside the ongoing extinction crisis."

Betsy Carpenter, U.S. News & World Report, *November 30, 1992*

"To attempt to preserve every species is to try to end the process of evolution."

Charles Oliver, Reason, *April 1992*

Off the Florida coast, the manatees are in trouble. These aquatic mammals have been reduced to a total population of about two thousand individuals. In past centuries, manatees were hunted for their flesh, oil, and hides. More recently, manatee populations have declined due to water pollution and habitat loss, problems that affect many other species. However, manatees also face an unusual threat to their survival: Increasing numbers of manatees have become hit-and-run victims, their bodies sliced by the propellers of motorboats. Officials have established speed limits on waterways where manatees live, but boat use has increased rather than declined, and many powerboaters continue to speed through these waters regardless of the regulations. In 1994 alone, fifty manatees were killed in collisions with boats—a significant number for a population already so close to extinction.

The manatee is just one species among hundreds worldwide that are experiencing drastic population reductions. Pandas, elephants, cougars, eagles, and wolves have all been classified as endangered, as have less prominent species such as the black lace cactus, the pupfish, and the burying beetle. Since 1973 in the United States, species designated as endangered (in current danger of extinction) or threatened (potentially in danger of extinction) have been protected under the Endangered Species Act (ESA), which prohibits the killing of any endangered species and most threatened species. As of early 1995, the ESA included 730 domestic endangered species and subspecies, with another 197 classified as threatened. In addition, 3,700 unlisted species suspected of being endangered or threatened were still awaiting review.

More than 500 foreign and oceanic species were also protected under the ESA as of 1995 and therefore were subject to U.S. trade sanctions. Most of these species are also classified as endangered by international organizations such as the Convention on International Trade in Endangered Species (CITES). According to the Worldwatch

Institute, a nonprofit organization that researches global problems, in the early 1990s three-fourths of the world's bird species were endangered, as were more than two-thirds of the primates and nearly half of all turtle species. Many conservationists are alarmed at the number of species in decline. As Linda M. Rancourt writes in *National Parks*, some scientists "estimate that the Earth is nearing an extinction of species unequaled since that of the age of the dinosaurs. We may lose as much as 15 percent of the world's organisms over the next 30 years."

Other scientists, however, do not believe that the current species extinction rate is a cause for concern. "Perhaps 95 percent of the species that once existed no longer exist," biologist Norman D. Levine contends:

> What species preservers are trying to do is stop the clock. It cannot and should not be done. Extinction is an inevitable fact of evolution. New species continually arise, and they are better adapted to their environment than those that have died out.

Since the beginning of European contact with North America in the sixteenth century, more than 500 North American species of plants and animals have become extinct. Those researchers who see extinction as a normal process argue that the United States has not greatly suffered economically or biologically from the loss of those species. In fact, Levine maintains, some extinctions may even be beneficial to humans: "Would it improve the Earth if even half of the species that have died out were to return? . . . The smallpox virus has been eliminated, except for a few strains in medical laboratories. Should it be brought back?" Writing in the *Freeman*, lawyer James A. Maccaro charges that human interference with the natural process of extinction may cause more harm than good. As an example, he cites wildlife officials' attempt to bolster kokanee salmon numbers by introducing shrimp into Montana rivers to serve as food for the salmon. Instead of helping the salmon population, the shrimp devoured large quantities of plankton, another important food source for the salmon. "As a result," Maccaro notes, "the number of salmon spawning in Lake McDonald in Glacier National Park fell from 100,000 to a mere 200." The drastic decline of salmon levels in the park adversely affected endangered bald eagle populations (which feed on the salmon) from northern Canada to the southwestern United States. Maccaro warns that "bureaucrats cannot tamper with [nature] without creating imbalances in the overall system."

Many environmentalists counter that human interference has already created an imbalance in nature—that is, an accelerated rate of species extinction. Current worldwide extinction rates are not caused by the normal process of evolution, they argue, but rather stem from human activities that endanger otherwise viable plant and animal species. "Dams bear much of the blame" in the decimation of salmon populations, Douglas H. Chadwick asserts in *National Geographic*, because they create "an obstacle course" for salmon attempting to swim upriver in order to spawn. Rancourt contends that "the popularity of rough, leathery alligator-skin handbags, shoes, and other clothing accessories" brought the American alligator to the edge of extinction. Numerous scientists consider the most potent human threat to other species to be the destruction of natural habitat, whether through logging old-growth forests, draining wetlands, converting prairies to farmland, or covering the coasts with housing developments. Geologist Warren D. Allmon argues that human actions are actually altering the process of evolution: Cod threatened by overfishing have begun to reproduce at an abnormally young age; mammals endangered by habitat loss are producing high numbers of albinos as a result of inbreeding. "The natural evolutionary processes that replace species are much slower than the elimination of species that we are currently causing," Allmon concludes.

While conceding that human activities may be endangering plant and animal species, many critics believe that human needs and concerns should be placed above those of species such as the kangaroo rat or the snail darter. According to Putting People First, an organization that promotes human development of natural resources, measures such as prohibiting logging in the Pacific Northwest forests may or may not save the spotted owl from extinction but will assuredly cost millions of dollars in lost revenues and deprive thousands of people of their jobs. Others object to the large amount of government spending on threatened and endangered species and claim that the funds would be better spent on human problems. Writing in *21st Century Science & Technology*, William O. Briggs Jr. points out that

> for the year 1990, taxpayers spent more than $90 million on 59 species—enough money to provide the average weekly unemployment compensation for 13,000 workers for a full year. The spotted owl cost taxpayers $9.2 million—enough money to provide 306 families with incomes of $30,000.

Summing up the opinion of many Americans, retired logger Kaye Kelso maintains that politicians "are going to have to change some of these laws that pertain to endangered species . . . to protect people instead."

The controversy surrounding endangered species primarily results from attempts to balance priorities: the importance of saving a given species versus the financial cost, the necessity of building a dam versus the impact on the local wildlife, the human desire for modern cities and technology versus the human love for pure and unfettered nature. The authors in *Endangered Species: Opposing Viewpoints* discuss the following questions: Is Extinction a Serious Problem? Can Endangered Species Be Preserved? Should Endangered Species Take Priority over Jobs, Development, and Property Rights? What Should Be U.S. Policy on Endangered Species in Other Countries? Are Humans an Endangered Species? These chapters examine the environmental, economic, and ethical debates that may determine the survival or extinction of the earth's endangered species.

CHAPTER 1
Is Extinction a Serious Problem?

Endangered Species

Chapter Preface

In 1813, American naturalist John Audubon observed one of the largest recorded flocks of passenger pigeons. Audubon, who estimated the flock as numbering over a billion; wrote that "the light of the noonday sun was obscured, as if by an eclipse." He also described the thousands of pigeons that were shot that day by local hunters. Like most people of the time, Audubon believed that human activities could not greatly affect the abundant population of passenger pigeons. By 1914, however, the passenger pigeon had become extinct, a victim of overhunting and the clearing of the eastern forest that had been its habitat.

Several decades later, in 1971, scientist David Etnier discovered the snail darter, a fish that lived close to the construction site of the Tellico dam. The snail darter, which conservationists believed only existed in a small population near the dam site, was declared an endangered species, and construction on the multimillion-dollar dam was halted by court order. Many people were outraged over the cessation of the dam project, and Congress eventually passed a special measure allowing the dam to be finished. Meanwhile, biologists had begun to find other snail darter populations outside of the region affected by the Tellico dam. Once evidence was found that the snail darter's numbers were greater than had previously been thought, its classification was upgraded from endangered to threatened.

The disparate fates of the passenger pigeon and the Tellico snail darter illustrate the difficulty of determining which species are endangered and which are thriving. Current research concerning endangered species is fraught with uncertainty over the significance of population numbers, environmental degradation, and human interference. Scientists disagree over the population estimates of particular species as well as the overall rate of worldwide extinctions.

More than 1.4 million existing species have been discovered and catalogued, but estimates of the total number of species range widely, from a conservative 2 million to as great as 100 million. Accordingly, estimates of the rate of species extinction vary dramatically. Many scientists agree with ecologist Stephen R. Edwards's contention that "documented extinctions peaked in the 1930s and . . . the number of extinctions has been declining since then." Others concur with Stanford biologists Paul and Anne Ehrlich that an "epidemic of extinctions" is currently under way. The authors in the following chapter debate whether the rate of extinction constitutes a serious problem.

VIEWPOINT 1

"Extinction is proceeding at a rapid rate, far above prehuman levels."

Extinction Is a Serious Problem

Edward O. Wilson

With or without human interference, extinction has always occurred: The dinosaurs disappeared sixty-five million years ago, several million years before humans walked the earth. However, some scientists, including Edward O. Wilson, believe that humans have now increased the pace of extinction far beyond natural levels. Wilson argues that species are becoming extinct at rates one thousand to ten thousand times above normal. This accelerated rate of extinction is largely due to humanity's overconsumption of the planet's resources, Wilson contends. A professor and curator at Harvard University, Wilson is the author of *The Diversity of Life*, from which this viewpoint is taken.

As you read, consider the following questions:
1. According to Wilson, how do scientists determine if a species is truly extinct?
2. What is the difference between extinction by "rifle shot" and by "holocaust," as defined by the author?
3. In Wilson's opinion, in what ways is humanity ecologically abnormal?

Reprinted by permission of the publishers from *The Diversity of Life* by Edward O. Wilson, Cambridge, Mass.: The Belknap Press of Harvard University Press, ©1992 by Edward O. Wilson.

Extinction is the most obscure and local of all biological processes. We don't see the last butterfly of its species snatched from the air by a bird or the last orchid of a certain kind killed by the collapse of its supporting tree in some distant mountain forest. We hear that a certain animal or plant is on the edge, perhaps already gone. We return to the last known locality to search, and when no individuals are encountered there year after year we pronounce the species extinct. But hope lingers on. Someone flying a light plane over Louisiana swamps thinks he sees a few ivory-billed woodpeckers start up and glide back down into the foliage. "I'm pretty sure they were ivorybills, not pileated woodpeckers. Saw the white double stripes on the back and the wing bands plain as day." A Bachman's warbler is heard singing somewhere, maybe. A hunter swears he has seen Tasmanian wolves in the scrub forest of Western Australia, but it is probably all fantasy.

In order to know that a given species is truly extinct, you have to know it well, including its exact distribution and favored habitats. You have to look long and hard without result. But we do not know the vast majority of species of organisms well; we have yet to anoint so many as 90 percent of them with scientific names. So biologists agree that it is not possible to give the exact number of species going extinct; we usually turn palms up and say the number is very large. But we can do better than that. Let me start with a generalization: *in the small minority of groups of plants and animals that are well known, extinction is proceeding at a rapid rate, far above prehuman levels. In many cases the level is calamitous: the entire group is threatened.*

Extinction in Process

To illustrate this principle, I will present a few anecdotes, out of many available: whenever we can focus clearly, we usually see extinction in progress. . . . Here are the examples:

- One fifth of the species of birds worldwide have been eliminated in the past two millennia, principally following human occupation of islands. Thus instead of 9,040 species alive today, there probably would have been about 11,000 species if left alone. According to a recent study by the International Council for Bird Preservation, 11 percent or 1,029 of the surviving species are endangered.

- A total of 164 bird species have been recorded from the Solomon Islands in the southwest Pacific. The International Union for Conservation of Nature and Natural Resources's *Red Data Book* lists only one as recently extinct. But in fact there have been no records for twelve others since 1953. Most of these are ground nesters especially vulnerable to predators. Solomon Islanders who know the birds best have stated that at

least some of the species were exterminated by imported cats.
- From the 1940s to the 1980s, population densities of migratory songbirds in the mid-Atlantic United States dropped 50 percent, and many species became locally extinct. One cause appears to be the accelerating destruction of the forests of the West Indies, Mexico, and Central and South America, the principal wintering grounds of many of the migrants. The fate of Bachman's warbler will probably befall other North American summer residents if the deforestation continues.

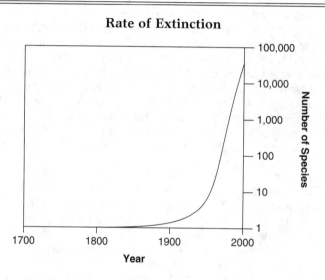

Source: Al Gore, *Earth in the Balance: Ecology and the Human Spirit*, 1992.

- About 20 percent of the world's freshwater fish species are either extinct or in a state of dangerous decline. The situation is approaching the critical stage in some tropical countries. A recent search for the 266 species of exclusively freshwater fishes of lowland peninsular Malaysia turned up only 122. Lake Lanao on the Philippine Island of Mindanao is famous among evolutionary biologists for the adaptive radiation of cyprinid fishes that occurred exclusively within the confines of the lake. As many as 18 endemic species in three genera were previously known; a recent search found only three species, representing one of the genera. The loss has been attributed to overfishing and competition from newly introduced fish species.
- The most catastrophic extinction episode of recent history may be the destruction of the cichlid fishes of Lake Victoria. . . .

From a single ancestral species 300 or more species emanated, filling almost all the major ecological niches of freshwater fishes. In 1959 British colonists introduced the Nile perch as a sport fish. This huge predator, which grows to nearly 2 meters in length, has drastically reduced the native fish population and extinguished some of the species. It is projected eventually to eliminate more than half of the endemics. The perch affects not only the fishes but the lake ecosystem as a whole. As the alga-feeding cichlids disappear, plant life blooms and decomposes, depleting oxygen in the deeper water and accelerating the decline of cichlids, crustaceans, and other forms of life. A task force of fish biologists observed in 1985, "Never before has man in a single ill advised step placed so many vertebrate species simultaneously at risk of extinction and also, in doing so, threatened a food resource and traditional way of life of riparian dwellers."

- The United States has the largest freshwater mollusk fauna in the world, especially rich in mussels and gill-breathing snails. These species have long been in a steep decline from the damming of rivers, pollution, and the introduction of alien mollusk and other aquatic animals. At least 12 mussel species are now extinct throughout their ranges, and 20 percent of the remainder are endangered. Even where extinction has not yet occurred, the extirpation of local populations is rampant. Lake Erie and the Ohio River system originally held dense populations of 78 different forms; now 19 are extinct and 29 are rare. Muscle [sic] Shoals, a stretch of the Tennessee River in Alabama, once held a fauna of 68 mussel species. Their shells were specialized for life in riffles or shoals, shallow streams with sandy gravel bottoms and rapid currents. When Wilson Dam was constructed in the early 1920s, impounding and deepening the water, 44 of the species were extinguished. In a parallel development, impoundment and pollution have combined to extinguish two genera and 30 species of gill-breathing snails in the Tennessee and nearby Coosa rivers.

Species with Narrow Habitats

- Freshwater and land mollusks are generally vulnerable to extinction because so many are specialized for life in narrow habitats and unable to move quickly from one place to another. The fate of the tree snails of Tahiti and Moorea illustrates the principle in chilling fashion. Comprising 11 species in the genera *Partula* and *Samoana*, a miniature adaptive radiation in one small place, the snails were recently exterminated by a single species of exotic carnivorous snail. It was folly in the grand manner, a pair of desperate mistakes by people in authority, which unfolded as follows. First, the giant African snail *Achatina fulica* was introduced to the islands as a food animal. Then, when it multiplied enough to become a pest, the carnivorous snail *Euglandina rosea* was introduced to control the *Achatina*. *Euglandina* itself multiplied prodigiously, advancing along a front at 1.2 kilometers a year. It consumed not only the giant African snail but every native tree snail along the way. The last of the wild tree snails became extinct on Moorea in 1987. On nearby Tahiti the same sequence is now unfolding. And in Hawaii the entire endemic tree-snail genus *Achatinella* is endangered by *Euglandina* and habitat destruction. Twenty-two species are extinct and the remaining 19 are endangered.

• A recent survey by the Center for Plant Conservation revealed that between 213 and 228 plant species, out of a total of about 20,000, are known to have become extinct in the United States. Another 680 species and subspecies are in danger of extinction by the year 2000. About three fourths of these forms occur in only five places: California, Florida, Hawaii, Puerto Rico, and Texas. The predicament of the most endangered species is epitomized by *Banara vanderbiltii*. By 1986 this small tree of the moist limestone forests of Puerto Rico was down to two plants growing on a farm near Bayamon. At the eleventh hour, cuttings were obtained and are now successfully growing in the Fairchild Tropical Garden in Miami.

• In western Germany, the former Federal Republic, 34 percent of 10,290 insect and other invertebrate species were classified as threatened or endangered in 1987. In Austria the figure was 22 percent of 9,694 invertebrate species, and in England 17 percent of 13,741 insect species.

• The fungi of western Europe appear to be in the midst of a mass extinction on at least a local scale. Intensive collecting in selected sites in Germany, Austria, and the Netherlands have revealed a 40 to 50 percent loss in species during the past sixty years. The main cause of the decline appears to be air pollution. Many of the vanished species are mycorrhizal fungi, symbiotic forms that enhance the absorption of nutrients by the root systems of plants. Ecologists have long wondered what would happen to land ecosystems if these fungi were removed, and we will soon find out.

Species on the Brink

For species on the brink, from birds to fungi, the end can come in two ways. Many, like the Moorean tree snails, are taken out by the metaphorical equivalent of a rifle shot—they are erased but the ecosystem from which they are removed is left intact. Others are destroyed by a holocaust, in which the entire ecosystem perishes.

The distinction between rifle shots and holocausts has special merit in considering the case of the spotted owl (*Strix occiden-*

talis) of the United States, an endangered form that has been the object of intense national controversy since 1988. Each pair of owls requires about 3 to 8 square kilometers of coniferous forest more than 250 years old. Only this habitat can provide the birds with both enough large hollow trees for nesting and an expanse of open understory for the effective hunting of mice and other small mammals. Within the range of the spotted owl in western Oregon and Washington, the suitable habitat is largely confined to twelve national forests. The controversy was engaged first within the U.S. Forest Service and then the public at large. It was ultimately between loggers, who wanted to continue cutting the primeval forest, and environmentalists determined to protect an endangered species. The major local industry around the owl's range was affected, the financial stakes were high, and the confrontation was emotional. Said the loggers: "Are we really expected to sacrifice thousands of jobs for a handful of birds?" Said the environmentalists: "Must we deprive future generations of a race of birds for a few more years of timber yield?"

Extinction Studies

A minority of scientists believes that a major extinction crisis is not at hand. They complain that extinction studies, often conducted on islands, exaggerate the losses because many continental species can escape into bordering areas once their prime habitat is removed. . . .

Skeptics point out that only two forest birds were lost as a result of the deforestation of the Eastern United States. But other scientists say tropical forests contain substantially more species than did the Eastern United States, and many cannot survive outside small ranges.

Another example offered by skeptics is Puerto Rico, a tropical island that lost most of its virgin forests at the beginning of the century and did not suffer massive extinctions. But the island is regularly buffeted by hurricanes and its species have evolved to adapt to tremendous disturbances.

Maura Dolan, *Los Angeles Times*, May 22, 1992.

Overlooked in the clamor was the fate of an entire habitat, the old-growth coniferous forest, with thousands of other species of plants, animals, and microorganisms, the great majority unstudied and unclassified. Among them are three rare amphibian species, the tailed frog and the Del Norte and Olympic salamanders. Also present is the western yew, *Taxus brevifolia*, source of taxol, one of the most potent anticancer substances ever found. The debate should be framed another way: what else awaits discovery in the old-growth forests of the Pacific Northwest?

Humans Are at Fault

The cutting of primeval forest and other disasters, fueled by the demands of growing human populations, are the overriding threat to biological diversity everywhere. But even the data that led to this conclusion, coming as they do mainly from vertebrates and plants, understate the case. The large, conspicuous organisms are the ones most susceptible to rifle shots, to overkill and the introduction of competing organisms. They are of the greatest immediate importance to man and receive the greater part of his malign attention. People hunt deer and pigeons rather than sowbugs and spiders. They cut roads into a forest to harvest Douglas fir, not mosses and fungi. . . .

Human demographic success has brought the world to this crisis of biodiversity. Human beings—mammals of the 50-kilogram weight class and members of a group, the primates, otherwise noted for scarcity—have become a hundred times more numerous than any other land animal of comparable size in the history of life. By every conceivable measure, humanity is ecologically abnormal. Our species appropriates between 20 and 40 percent of the solar energy captured in organic material by land plants. There is no way that we can draw upon the resources of the planet to such a degree without drastically reducing the state of most other species. . . .

If past species have lived on the order of a million years in the absence of human interference, a common figure for some groups documented in the fossil record, it follows that the normal "background" extinction rate is about one species per one million species a year. Human activity has increased extinction between 1,000 and 10,000 times over this level in the rain forest by reduction in area alone. Clearly we are in the midst of one of the great extinction spasms of geological history.

VIEWPOINT 2

"Without extinction, without a loss of current variety, future variation diminishes."

Extinction Is Not a Serious Problem

Julian L. Simon and Aaron Wildavsky

Julian L. Simon, a professor of business administration at the University of Maryland, is the author of *Population Matters: People, Resources, Environment, and Immigration*. Aaron Wildavsky is a professor of political science and public policy at the University of California at Berkeley and the author of numerous articles. Simon and Wildavsky maintain that many of the forecasts of rapid species extinction are based on faulty reasoning, incomplete data, and guesswork. The extinction of species is a natural process that allows new species to evolve and thrive, the authors assert. Furthermore, they state, new scientific developments enable humans to conserve samples of endangered species while allowing new species to arise.

As you read, consider the following questions:
1. What objections do the authors make to the current definitions of species?
2. What do the authors believe are the benefits of extinction?
3. How do conservationists hinder policymaking concerning endangered species, according to Simon and Wildavsky?

Reprinted by permission of Transaction Publishers from Julian L. Simon and Aaron Wildavsky, "Species Loss Revisited," *Society*, November/December 1992. Copyright ©1992 by Transaction Publishers; all rights reserved.

It is proper to be concerned about possible dangers to species. Individual species, and perhaps all species taken together, constitute a valuable endowment, and we should guard their survival just as we guard our other physical and social assets. But we should strive for as clear and unbiased an understanding as possible of species loss in order to make the best possible judgments about how much time and money to spend in guarding them, in a world in which this valuable activity must compete with other valuable activities, including the guarding of human life.

The importance of the topic is clear from the far-reaching extent of the policies suggested. Edward O. Wilson and Paul Ehrlich actually ask that governments act "to reduce the scale of human activities." More specifically, they want us "to cease 'developing' any more relatively undisturbed land," because "every new shopping center built in the California chaparral and every swamp converted into a rice paddy or shrimp farm means less biodiversity." Though this is not necessarily pro-species survival, which depends on many other factors, the views expressed by them are certainly anti-growth.

Defining Species

Before discussing rates of extinction, we must touch on an issue that complicates such estimates—the definition of a species. Referring to the "never-ending arguments about the definition of the species category," Ernst Mayr infers that "those who do not work with species but with cells or molecules may think that the species is an arbitrary and insignificant concept in biology." He argues otherwise. "What, then," Mayr asks, "is biological classification? Unhappily," he concludes, "no agreement on the answer to this question exists yet among biologists." He then argues that the taxonomy is based not on the similarity of the species but on their common ancestors.

The *Encyclopaedia Britannica* defines species as "groups of individuals that resemble one another more than they resemble any others." But what constitutes "resemblance"? If we adopt the definition of "who mates with whom," we will reduce the number of species; if we give the more general definition that "species are groups of organisms sharing many traits, or characteristics in common," the classifiers have a lot of room to raise or lower the number. If we use common DNA as a criterion, we suspect the number of species would greatly diminish. The basic forecast for loss of species comes from Thomas Lovejoy in the *Global 2000 Report*:

> What then is a reasonable estimate of global extinctions by 2000? Given the amount of tropical forest already lost (which is important but often ignored), the extinctions can be estimated. In the low deforestation case, approximately 15 percent of the planet's species can be expected to be lost. In the

high deforestation case, perhaps as much as 20 percent will be lost. This means that of the 3 to 10 million species now present on the earth, at least 500,000 to 600,000 will be extinguished during the next two decades.

This extract summarizes a table of Lovejoy's that shows a range of between 437,000 and 1,875,000 extinctions out of a present estimated total of 3 to 10 million species. The table in turn is based on a linear relationship running from zero percent species extinguished at zero percent tropical forest cleared, to about 95 percent extinguished at 100 percent tropical forest clearing. The main source of differences in the range of estimated losses is the range of 3 to 10 million species in the overall estimate. The basis of any useful projection must be some body of experience collected under some range of conditions that encompass the expected conditions, or that can reasonably be extrapolated to the expected conditions.

But none of Lovejoy's references contains a scientifically impressive body of experience. . . .

The Island Model

No one knows how many species actually exist. As recently as the 1960s, scientists thought there were about 4 million species. Then estimates exploded when biologists realized how numerous and diverse life was in tropical rain forests. Estimates of global species counts range as high as 100 million, but taxonomists have catalogued only a little fewer than 1.4 million. The rest is guesswork.

Almost every estimate of species loss is based on computer models. These models rely upon assumptions that overstate potential extinction rates. Modelers assume that habitats are like islands which shrink as development spreads. But the analogy is faulty. On islands, animals can't adapt to rising seas, but many animals can adapt to human development, especially when the development in question is light farming or low-density housing. Another flaw is that the computer models are based on thirty-year-old tropical-island research. Since tropical areas have more life per square foot than temperate areas, habitat loss is bound to take a larger toll in the tropics. By extrapolating from tropical data, some models overstate the human impact on wildlife.

Richard Miniter, *National Review*, July 6, 1992.

Confirmation of the absence of scientific evidence for rapid species extinction is implicit in the nature of the "evidence" cited by Edward O. Wilson. He says that "the extinction problem" is "absolutely undeniable." All he cites are "literally hundreds of anecdotal reports." The very reason for the scientific method in estimating rates is that anecdotal reports are of little or no value, and often mislead the public and policymakers; that is why expensive censuses and other data-gathering instruments are mounted. . . .

Good Effects of Extinction

In the case of species extinction, as with many other public issues, there is a tendency in both technical discussion and in the press to focus only upon the bad effects, and to exclude from consideration possible good effects. Thus the discussion typically includes numbers and lists of extinguished species such as birds. Typically, no mention is made of the fact that there are concurrent opposite effects on species. Two of these effects became better understood in the 1980s: the species-promotion effects of less-than-catastrophic perturbation in general, and the species-promotion effects of human activities, including the isolation of species.

For example, Ariel Lugo notes that because humans have facilitated immigration of species and created new environments, exotic species have successfully become established in the Caribbean islands. This has resulted in a general increase in the total inventories of bird and tree species. In tropical Puerto Rico, where "human activity reduced the area of primary forests by 99 percent," as great a reduction as could be imagined, "seven bird species became extinct after 500 years of human pressure and exotic [newly resident] species enlarged the species pool. More land birds have been present on the island in the 1980s (ninety-seven species) than were present in pre-Colombian time (sixty species)."

Perhaps, says Lugo, conservation biologists make mention of the extinctions but not of the newly resident species because "there is a clear aversion to exotic [newly resident] species by preservationists and biologists (in cases such as predatory mammals and pests, with good reason!)." Such an aversion to new species may have something to do with the idea that humankind is somehow artificial and not "natural." Consider the language of Norman Myers, who has played as important a role as anybody in sounding the alarm about species extinction. He says: "[W]hereas past extinctions have occurred by virtue of natural processes, today the virtually exclusive cause is man." If it is species variety that is at issue, as is usually claimed, rather than preserving the species as they exist today, then new species should count as much as old.

Maintaining the Amazon river basin and other areas in a state of stability, according to a recent body of research, might even be counterproductive for species diversity. Natural disturbances, as long as they are not catastrophic, may lead to discontinuity in environments and to consequent isolation of species that may "facilitate ever-increasing divergence." Paul Colinvaux goes on to suggest that the highest species richness will be found not where the climate is stable but rather where environmental disturbance is frequent but not excessive. The same line of thought leads to possible benefits from interventions by humankind.

Determining the Rate of Extinction

During the 1980s, there was increasing recognition that the rate of species loss really is not known. "Regrettably," Myers now writes, "we have no way of knowing the actual current rate of extinction in tropical forests, nor can we even make an accurate guess." And Colinvaux refers to other extinctions as "incalculable." One would think that this state of affairs would make anyone leery about estimating future extinctions. Nevertheless, Myers continues, "we can make substantive assessments by looking at species numbers before deforestation and by applying the analytical techniques of biogeography. . . . According to the theory of island biogeography, we can realistically reckon that when a habitat has lost 90 percent of its extent, it has lost half of its species."

Biogeography is mere speculation, inconsistent with Lugo's finding that in Puerto Rico, the "massive forest conversion did not lead to a correspondingly massive species extinction, certainly nowhere near the 50 percent alluded to by Myers." Some conservationists have become frustrated at their inability to document rapid species extinction that would justify calls for government regulation; they also are annoyed at our writing about the actual state of the evidence. As a prominent conservationist, Jared Diamond, responded to our article, "documenting degree of threat is often difficult, and economists and others who wish to downplay the risk of an extinction crisis can easily dispute this case or that case, casting doubt even on the claim that 5 percent of the world's birds are threatened." Because of the drastic measures proposed to limit species extinction, we think the burden of proof must rest with those who make these policy-relevant claims. Otherwise, anyone could justify policy changes on only their say-so. Diamond therefore has suggested looking at the evidential issue in an entirely different fashion, one that is quite out of keeping with ordinary scientific practice. Normally, he writes,

> [S]pecies are to be considered extant until proven extinct. [But] for most species of the tropics or other remote regions, that is, for most of the world's species a more appropriate assumption would be "extinct unless proven extant." We biologists should not bear the burden of proof to convince economists advocating unlimited human growth [an inaccurate description of one of the authors of this viewpoint] that the extinction crisis is

real. Instead, it should be left to those economists to fund research in the jungles that would positively support their implausible claim of a healthy biological world.

Reversing the Burden of Proof

This is a "reversal of proof burden," as David Western puts it. It implies that it is enough that a warning be sounded and a charge be made to cause the community to proceed as if the case has been proved. By analogy, if someone says that the forest floor has turned to blue cheese and advocates that the government should immediately begin to package and sell the cheese, those who question this policy have the responsibility of demonstrating that the blue-cheese transformation has not taken place. This intellectual strategy suggests that the biologists now despair of making their case with the usual tools of scientific inquiry and ask instead for support on the basis of non-evidential faith.

To go one step further: the conservationists premise their forecast of rapid species extinction on there being a rapid rate of deforestation now and in the future. Even if the rate of deforestation were indeed rapid, there is little or no evidence that would justify inferring a rate of species extinction of Lovejoy's projected magnitude. But this line of argument is rendered even weaker by the fact that the historical evidence does not support their projections of deforestation.

The Importance of Extinction

Should everything alive be perpetuated under all conditions? Clearly not. Without extinction, without a loss of current variety, future variation diminishes. Prohibiting extinction implies that existing species should be protected against future species that will not have the opportunity to emerge. In earlier times, species could not be preserved without diminishing the prospects for new variety. But nowadays seed banks can store species and recombinant DNA enables the generation of variants on old forms. To a growing extent, humankind can simultaneously retain the old and generate new variety without needing to trade one for the other.

It matters greatly whether human beings are considered part of nature or not. If humankind is not considered part of nature—indeed, is antithetical to it—then people are perceived as threats to natural variety acting to crowd out existing species. But if humankind is considered to be part of the natural order, then it is not only human beings who must adapt to other species, but those nonhuman forms of life that must also adapt to us.

Another difficulty is that the goals of those who worry about species diversity are amorphous. Sometimes they emphasize the supposed economic benefits of species diversity. Other writers, like Quinn and Hastings, say that "maximizing total species diversity is rarely if ever the principal objective of conservation strategies. Other aesthetic, resource preservation, and recreational values are often more important." And Lovejoy says, most inclusively:

> What I'm talking about is rather the elusive goal of defining the minimum size [of habitat] needed to maintain the characteristic diversity of an ecosystem over time. In other words, I think the goals of conservation aren't simply to protect the full array of plant and animal species on the planet, but rather also to protect them in their natural associations so that the relationships between species are preserved and the evolutionary and ecological processes are protected.

Such a lack of agreement on what the goals should be makes it difficult to compare the worth of a species-saving activity against another value. For example, what are the relative worths of maintaining the habitat on Mount Graham, Arizona, for about 150 red squirrels who could be kept alive as a species elsewhere versus using twenty-four acres for an observatory that would be at the forefront of astronomical science? There is much less basis here for a reasoned judgment in terms of costs and benefits than even with such thorny issues as electricity from nuclear power versus from coal, or decisions about supporting additional research on cancer versus using the funds for higher social security payments or for defense or even for lower taxes.

Policymaking is also made difficult by conservationists asserting on the one hand that the purpose of conserving is that it is good for human existence, and on the other hand asserting that human existence must be limited or reduced because it is bad for the other species. For example, Diamond writes that there are many realistic ways we can avoid extinctions, such as by preserving natural habitats and limiting human population growth. But he also urges that humans should preserve the species because humans need them for existence!

Blaming Humans

The view that the interests of humans and of other species are opposed leads to humankind being seen in a rather ugly light. "[O]ur species has a knack for exterminating others, and we've become better killers all the time." A recent article is entitled "Extinction on Islands: Man as a Catastrophe."

It is quite clear that species are seen by many as having value quite apart from any place they play in human life, a value that is seen as competitive with the value of human life. "Although human beings are biologically only one of the millions of species that exist on Earth," Peter Raven writes, "we control a highly disproportionate share of the world's resources." Is it unfair that we humans "control" more resources than do eagles, mosquitoes, or the AIDS virus?

Verifying Extinction

Since the vast majority of species are unknown (Dr. Edward O. Wilson asserts that it would take the lifetime work of 25,000 scientists to catalog another 1 million), it is impossible to say with certainty how many have or will become extinct. Indeed, no one has ever actually witnessed the demise of a species. The standard for "extinction" is simply man's inability to find specimens of a given species over a given period of time. A number of species classified as "extinct" have subsequently reappeared. It was a tenet of evolutionary faith, for example, that the coelacanth, one of a group of lobe-finned fish from which four-legged land animals were said to have developed, had been extinct for hundreds of millions of years. But in 1939, a living coelacanth was caught off the coast of South Africa, and others have since been found near Madagascar.

Robert W. Lee, *The New American*, July 27, 1992.

Biologists with whom we have discussed this material agree that the numbers in question are most uncertain. But, they also say that the numbers do not matter scientifically. The conclusion would be the same, they say, if the numbers were different even by several orders of magnitude. But if so, why mention any numbers at all? The answer, quite clearly, is that these numbers do matter in one important way: they have the power to frighten in a fashion that smaller numbers would not. We can find no scientific justification for such use of numbers. . . .

Informed Decisions

The question remains: How should decisions be made, and sound policies formulated, with respect to the danger of species extinction? It certainly makes no sense to attempt to save all species at any cost, any more than to attempt to save all human lives at any cost. Society must establish some informed estimates about the present and future social value of species, just as it is necessary to estimate the value of human life to make rational policies about public health care services, such as hospitals and surgery, and about indemnities to accident survivors. And just as with human life, valuing species relative to other social goods will not be easy, especially because we must place some value on some species that we do not know about. But the job must be done in the best possible fashion. . . .

Perhaps we should look backward and ask, what were the species extinguished when the settlers cleared the Middle West of the United States? Are we the poorer now for their loss? Obviously, we cannot know in any scientific way. But can we even imagine that we would be enormously better off with the persistence of any hypothetical species?

At a time when there appear frequent reports on the extraordinary possibilities of genetic engineering, for example, "Animals Altered to Produce Medicine in Milk; Scientists Say Rare Drugs Could Be Manufactured with Relative Ease," it is beginning to seem ludicrous to justify extraordinary expense for protecting an animal like the grey squirrel, which may not even be genetically distinct, on the grounds that its gene plasm will be valuable for human life in the future. . . .

What can one say with any degree of confidence about what variety consists of, how to measure it, why and in what way it is valuable and to whom? At present, next to nothing. No *prima facie* case for expensive policy for safeguarding species can be made without more extensive analysis. The question certainly deserves deeper thought and more careful and wide-ranging analysis than has been done until now.

3 VIEWPOINT

"The degradation of whole ecosystems . . . [is] the single most important factor behind the current mass extinction of species."

Deforestation Decreases Biodiversity

John C. Ryan

Many scientists believe that the destruction of ecosystems is a major contributor to the loss of biodiversity. One ecosystem of great concern is the tropical rain forest, which contains an estimated fifty percent of the world's plant and animal species. In the following viewpoint, John C. Ryan argues that unchecked deforestation in tropical and temperate rain forests presents a great threat to numerous species. Loss of these species will result in a dangerous depletion of genetic variety, he warns. Ryan is a research associate specializing in forests, biodiversity, and environmental history at the Worldwatch Institute in Washington, D.C., a nonprofit research organization that focuses on problems of global concern.

As you read, consider the following questions:

1. According to Ryan, on what three levels is biodiversity analyzed?
2. How are tropical forests important to rural people, in the author's view?
3. What does Ryan mean by the term "New Forestry"?

From John C. Ryan, "Conserving Biological Diversity," *State of the World 1992*. Copyright 1992 by The Worldwatch Institute. Reprinted by permission.

Biological diversity—complex beyond understanding and valuable beyond measure—is the total variety of life on earth. No one knows, even to the nearest order of magnitude, how many life forms humanity shares the planet with: Roughly 1.4 million species have been identified, but scientists now believe the total number is between 10 million and 80 million. Most of these are small animals, such as insects and mollusks, in little-explored environments such as the tropical forest canopy or the ocean floor. But nature retains its mystery in familiar places as well—even a handful of soil from the eastern United States is likely to contain many species unknown to science.

Mass Extinction

Despite the vast gaps in knowledge, it is clear that biodiversity—the ecosystems, species, and genes that together make life on earth both pleasant and possible—is collapsing at nothing less than mind-boggling rates. Difficult as it is to accept, mass extinction has already begun, and the world is irrevocably committed to many further losses. Harvard biologist Edward O. Wilson estimates that at a minimum, 50,000 invertebrate species per year—nearly 140 each day—are condemned to extinction by the destruction of their tropical rainforest habitat. Large creatures as well as small are vanishing: deforestation condemns at least one species of bird, mammal, or plant to extinction daily.

Moreover, biological impoverishment is occurring all over the globe. Ecosystems with fewer species than rainforests have, such as islands and freshwater lakes, are probably losing even greater proportions of their varied life forms. Genetic varieties within species and entire natural communities are also disappearing, likely at rates greater than the extinction of species themselves.

Protection of wildlands will be the top priority of any meaningful strategy to safeguard the world's biological heritage. True protection of these ecosystems alone will require sweeping changes in the way humanity views and uses land and a commitment to limit the amount of the earth's bounty that society appropriates to itself. But in order to staunch the massive bleeding of life from the planet, humanity must learn not only to save diversity in remote corners of the world but also to maintain and restore it in the forests and waters that we use, and in the villages and cities where we live.

Why should disappearing beetles, plants, or birds concern us? To biologists, and to many others, the question hardly needs asking: A species is the unique and irreplaceable product of millions of years of evolution, a thing of value for scientific study, for its beauty, and for itself. For many people, however, a more compelling reason to conserve biological diversity is likely to be pure self-interest: Like every species, ours is intimately dependent on others for its well-being.

Time after time, creatures thought useless or harmful are found to play crucial roles in natural systems. Predators driven to extinction no longer keep populations of rodents or insects in check; earthworms or termites killed by pesticides no longer aerate soils; mangroves cut for firewood no longer protect coastlines from erosion. Diversity is of fundamental importance to all ecosystems and all economies.

The Degradation of Ecosystems

Biodiversity is commonly analyzed at three levels: the variety of communities and ecosystems within which organisms live and evolve, the variety of species, and the genetic variety within those species themselves. The degradation of whole ecosystems, such as forests, wetlands, and coastal waters, is in itself a major loss of biodiversity and the single most important factor behind the current mass extinction of species.

Home to at least half the planet's species, tropical forests have been reduced by nearly half their original area, and in 1990 deforestation claimed 17 million hectares (one hectare equals 2.471 acres), an area about the size of Washington state. In Benin, Côte d'Ivoire, Western Ecuador, El Salvador, Ghana, Haiti, Nigeria, and Togo, forests have all but disappeared. In most nations, forests occur increasingly in small fragments surrounded by degraded land, with their ability to sustain viable populations of wildlife and vital ecological processes impaired.

Brazil has more tropical forest—and likely more species—than any other nation. Massive deforestation continues there, but it has slowed appreciably since its 1987 peak, thanks to unusually rainy weather, changes in government policy, and a slowdown in the Brazilian economy. Moreover, with nearly 90 percent of its groves still standing, by national or international standards the Brazilian Amazon is relatively untouched. Brazil's most endangered ecosystems are its unique coastal forests. Logging and agricultural and urban expansion have destroyed more than 95 percent of the once-vast Atlantic coastal rainforests and the coniferous Araucaria forests of southern Brazil.

Outside the tropics, a number of ecosystem types have been all but eliminated from the planet, including the tallgrass prairies of North America, the great cedar groves of Lebanon, and the old-growth hardwood forests of Europe and North America. Less widespread than their tropical counterparts, temperate rainforests are probably the more endangered ecosystem. Of the 31 million hectares once found on earth, 56 percent have been logged or cleared. In the contiguous United States, less than 10 percent of old-growth rainforests survive, scattered in small fragments throughout the Pacific Northwest. In the rainforests of British Columbia, only one of 25 large coastal watersheds has wholly escaped logging. . . .

The most familiar type of biodiversity loss is the decline of species, a process now occurring at thousands of times its natural "background" rate. The majority of species—and of extinctions—are invertebrates of the tropical forest too numerous to identify, let alone monitor the status of. Outside the tropics, the situation is somewhat easier to track. All 41 species of Hawaiian tree snail, for example, were listed as endangered by the U.S. government in 1981; today only two remain in substantial numbers, and they are declining rapidly. . . .

Reprinted by permission: Tribune Media Services.

Paralleling patterns of animal diversity, two-thirds of the world's plants are found in the tropics. Although prehistorical extinction spasms tended to claim mostly animals, plants are now threatened with extinction on a large scale. Peter Raven of the Missouri Botanical Garden estimates that one-fourth of all tropical plants are likely to be wiped out in the next 30 years. . . .

For the past century, nature conservation efforts have focused on the protection of habitats in parks and other reserves. This strategy has had an important role in preserving biological diversity. Today there are just under 7,000 nationally protected areas in the world, covering some 651 million hectares, or about 4.9 percent of the earth's land surface.

Several nations have, on paper at least, set impressive proportions of their territory off-limits to development: Bhutan, Botswana, Czechoslovakia, Panama, and Venezuela notably have over 15 percent of their lands designated as parks. National or global figures, however, mask great unevenness. Parks in Chile, for example, are concentrated high in the scenic Andes, and more than half of Chile's unique vegetation types are not protected at all.

Globally, high-altitude habitats have received a disproportionate share of protective efforts, while others of greater biological significance (such as lowland forests, wetlands, and most aquatic ecosystems) have been neglected.

Fading Knowledge

Today both modern and indigenous conservation systems are unraveling. In many areas, such as the Amazon basin, the often intimate knowledge of nature possessed by indigenous people is fading even faster than nature itself. On average, one Amazon tribe has disappeared each year since 1900. Especially for medicinal plants, traditional crops, and other life forms favored and used by native people, acculturation and the loss of traditional management systems are among the greatest threats to biological conservation.

The relationship between cultural diversity and biological diversity is more ambiguous in the case of large animal species. But the indigenous practices that do conserve wildlife are also vanishing as native cultures and territories succumb to the expanding influence of the global commercial economy. Orangutans, for example, have been virtually wiped out in the Malaysian state of Sarawak by destruction of their rainforest habitat and hunting. Only along the upper reaches of the Batang Ai River in southern Sarawak do they thrive, in part because local Iban people believe it is taboo to kill them. But the Iban, constantly told that their culture is backward, are abandoning their traditional beliefs, and orangutan hunting is reportedly on the increase. . . .

Tropical forests are critical to hundreds of millions of rural people as sources of nutrition, health care, raw materials, and cash income. Traditional medicine, based largely on tropical plants, nurtures four-fifths of humanity, while rainforest plants provide key ingredients in pharmaceuticals worth tens of billions of dollars annually. International commerce in the most widely traded nontimber forest product, rattan (palm stems used for wicker furniture and baskets), is alone worth roughly $3 billion annually.

As with rangelands, fisheries, and wildlife throughout the developing world, forests have long been managed as common property resources, often with few negative ecological impacts.

Many communal management systems are unraveling as populations surge, traditional cultures erode, and national governments confiscate or privatize resources held by communities. In addition, when subsistence-level economies have adopted modern technologies or become commercialized, increasing levels of production have tended to strain local ecosystems without improving local welfare.

Unraveling the Web

Researchers in the Amazon observed the web of life unravel after a rancher, at the scientists' request, allowed patches of rain forest to remain in an area that had been cleared in 1979 for cattle.

Pigs disappeared from some of the isolated patches, triggering the demise of seven species of frogs, according to one of the researchers from the Smithsonian Institution. The frogs needed the pigs' wallows to survive.

Some species of birds also disappeared because they were dependent on army ants. The patch of forest was too small to sustain enough swarms of the voracious ants, which scare up insects from among the leaves and under logs as they approach. The birds survive by feeding on the insects fleeing the ants.

Maura Dolan, *Los Angeles Times*, May 22, 1992.

From Southeast Asia's rattan to the fish and fruits of the Peruvian Amazon, the usual fate of species that gain long-term popularity in industrial markets is depletion. And the usual lot of their harvesters is continued poverty, as the profits from their work are siphoned off by powerful intermediaries and elites. Brazil nut gatherers, for example, receive about 4¢ a pound for their labors, just 2–3 percent of the New York wholesale price. Three-fourths of the market is controlled by three companies, owned by three cousins.

Impacts on the Food Chain

Rainforest trees form the base of a complex food chain. In turn, they rely on birds, insects, mammals, and even fish to disperse their pollen and seeds. According to ecologist Charles M. Peters of the New York Botanical Garden, "it is obvious that removing commercial quantities of fruits, nuts, and oil seeds will have an impact on local animal populations." The economic incentive to "enrich" forests with money-making species and cut down their competitors could also reduce the diversity of plant species (along with symbiotic animals) within extraction areas.

Without careful management, losses at the genetic level are also likely: as the best fruits or leaves from a given population are continually harvested, the "inferior" plants may be the ones to survive over time.

The dangers of market-oriented extractive economies can be minimized by carefully choosing species and ecosystems. Some tropical forest areas are naturally less diverse than others and thick with commercially promising species—such as the vitamin C-rich fruit of the *camu camu* that grows in dense stands throughout the floodplains of the Peruvian Amazon. Ancient yet seminatural forests also hold great promise. Anthropologist Christine Padoch of the New York Botanical Garden reports finding fruit-filled, remarkably diverse groves (44 tree species within 0.2 hectares) in Kalimantan, Indonesia, that had been created by generations of villagers casually planting, weeding, and even spitting out fruit seeds over their shoulders.

In some cases, making the use of diversity rewarding to local people can be aided by developing markets for new or unknown products. But in all cases, this approach will work only if the fundamental concerns of the rural poor are addressed first. Political reforms will be required to secure rural people's rights to land and resources, reduce their vulnerability to exploitation and violence by outsiders, and ensure that the benefits of conservation stay within local communities.

Reconciling diversity with development will require loosening the grip of industry on natural resources and accepting that production of any single commodity will be constrained if ecosystems are managed for diversity. The success of some U.S. foresters' efforts to develop a "New Forestry," for example, depends on whether they can free forest management from its historical timber bias. Rather than homogenize ecosystems in order to maximize wood production, New Forestry ostensibly aims to maintain and use the ecological complexity of natural forests. But because diversity conservation will require protecting the few scattered remnants of pristine forest and logging other areas less intensively, New Forestry cannot succeed without major reductions in the timber harvest on national forests—a politically charged issue.

Sustainable Nonuse

While New Forestry, extractive reserves, and other programs of sustainable commercial use of ecosystems are in their infancy, it is already clear that their success depends on programs of sustainable *nonuse*. Protecting large areas of undisturbed forest near logging or extraction areas is the only way to ensure the survival of animals that pollinate and disperse the seeds of commercially important tropical trees. Studies in the western United States have shown intact stands of native forest to harbor insect predators that safeguard plantations from pest outbreaks; the protection of coral reefs from fishing in Belize and the Philippines has revived local fishing economies by providing depleted fish stocks room to grow. Ultimately, only natural areas can tell us if our uses are in fact sustainable. As essayist Wendell Berry writes, "We cannot know what we are doing until we know what nature would be doing if we were doing nothing.". . .

Many of the changes needed for biodiversity's sake are a step beyond those considered in discussions of sustainable development. The human costs of a lost medicinal plant may never be known, and any economic deterioration that results from biological decline could take years or decades to appear. But the protection of biodiversity has to be considered a basic requirement of sustainability—passing on to future generations a world of undiminished options—and a fundamental moral responsibility as travelers on the only planet known to support life.

VIEWPOINT 4

"Any large rain forest should be able to withstand at least some clearing without any loss of species."

Deforestation's Effects on Biodiversity Are Exaggerated

Shawn Carlson

In the following viewpoint, Shawn Carlson insists that, while the clearing of the rain forests presents ecological problems, the threat to biodiversity has been greatly exaggerated. Carlson argues that the formula used to estimate the number of species and the rate of extinction in the rain forests is highly inaccurate, leading to disturbing but false statistics. In addition, he contends that these faulty percentages have led scientists and environmentalists to suggest alarmist measures that actually hinder attempts to save the rain forests. Carlson is a physicist at Lawrence Berkeley Laboratories and a technical consultant to the Committee for the Scientific Examination of Religion, a secular humanist organization.

As you read, consider the following questions:

1. Why does the island model fail to accurately reflect biodiversity in rain forests, in Carlson's opinion?
2. According to the author, what inaccuracies are found in the television ads of environmental groups?
3. What measures does Carlson recommend to save the rain forests?

"Death in the Rain Forests," by Shawn Carlson, *The Humanist*, March/April 1992, is reprinted with permission of the American Humanist Association, ©1992.

Close your eyes and think of paradise. What do you see? Lush warm tropical forests, perhaps? Exotic birds and a plethora of strange creatures busily going about the business of survival? For me, nothing could come closer to paradise than a rain forest—a place of fantastic diversity, unfathomably rich in color and form.

Unfortunately, there is one particularly efficient primate going about its business in the rain forests as well. If the relentless industry of *Homo sapiens* proceeds unabated, it will ultimately devour these forests, forever eradicating at least half of all life on Earth. Fortunately, the situation is not as bleak as some people insist. All the vital signs suggest that the world's rain forests are in serious—but *not* critical—condition.

Wet Deserts

Surprisingly, these lavish forests grow in what biologists call *wet deserts*—extremely poor soil whose nutrients are quickly leached away by over 40 inches of rain each year. This perpetual soaking leaves behind those minerals which do not dissolve in water and which, unhappily for the flora, cannot be absorbed by plant roots. Only about 0.1 percent of the nutrients released by the decaying matter on the forest floor ever feeds the next generation of vegetation. Furthermore, the struggle for sunlight favors the development of a network of high branches with broad expanses of large leaves to catch every morsel of the sun's nourishment. The branches of neighboring trees grow together into a thick canopy of leaves which absorbs rain water like green blotting paper and then slowly releases it to keep the air saturated with moisture. These canopies provide rich habitats for millions of organisms (mostly insects), which have evolved in such a way that they spend their entire lives in the tree tops and are therefore almost never seen by humans.

It is not surprising then that most of the rain forests' inhabitants have never been described by science. Just how many species do these forests harbor? Nobody knows—not even to within a factor of ten. Any estimate of biodiversity must be based upon surveys of areas which are small enough for biologists to personally scrutinize. To estimate the total number, the species count in these areas must somehow be extrapolated to the entire rain forest. Unfortunately, we have no model, no mathematical formula, which relates the number of species to rain-forest area. Still, assessing the damage now being wreaked on the forests requires some estimate to be made.

Although the species-area formula is not known for rain forests, it is well known for so-called islands—relatively small and isolated habitats. In this context, an "island" can be a conventional island—an outcropping of land in an ocean—but it can also be a mountaintop above the timber line or an oasis surrounded by desert. On an island, the total number of species scales roughly as the fourth root of the island's area. Thus, if two islands had the same environment but one was 16 times larger than the other, the larger island would be roughly twice as species rich as the smaller one. To estimate how many species make a particular island their home, a biologist could simply count all the species in some small but representative piece of the island and then extrapolate to the whole. Even if the species started out evenly distributed throughout the island, the destruction of habitat would put pressures on the surviving organisms. Put simply, any given area can only support so many species, and, if there are too many, competition for resources will drive some of them into extinction. But while the island model predicts that every acre cleared contributes to extinctions, keep in mind that a fourth-root dependence is very slow. The island model predicts that annihilating 50 percent of a habitat extinguishes only 16 percent of its inhabitants. If 99 percent of the island is destroyed, 32 percent of the species should still survive in the 1 percent remaining.

Problems with the Island Model

Since there are no good models for rain forests, most people use the island model to estimate both rain-forest biodiversity and extinction rates. But since rain forests are not very much like islands, the island model can give only the crudest estimates for the rain forests' problems.

Just how badly does the island model do? Islands can be rather sparse in species, but rain forests absolutely *brim* with biodiversity. It is not uncommon, for example, for a single leguminous tree in Peru to be home to more than 40 species of ants representing more than 25 genera. That's comparable to the total ant diversity found in the whole of Brittany! Most islands just don't compare.

Also, rain forests are usually *much* bigger than islands. They occupy gaping swaths of South America, Africa, and Indonesia and cover over 60 countries. Biologists do try to break up the rain forests into a number of smaller microclimates and use the island model to estimate the total number of species of each. Even so, the survey areas must be scaled to regions that are sometimes *millions* of times larger. In fact, the island model almost certainly fails for these huge extrapolations because, while new ecological niches (the engines which drive biodiversity) are easy to create on small islands, they are not so easily created in huge forests. A small island contains comparatively few niches. Adding area to an island adds lots of new niches and therefore enhances biodiversity, but adding area to a large forest tends only to duplicate the niches that are already available. And, according to University of Maryland rain-forest expert Patrick Kangas, this is what the rain-forest data actually show. In a recent issue of *Science*, he said, "There's a finite number of species within any community type. As you continue to move out, the number levels off." In other words, the number of species in a rain forest rises with area, but only to a point. And that's good news because it means that *any large rain forest should be able to withstand at least some clearing without any loss of species.*

©1992. Reprinted by permission of Mike Keefe.

And there's more good news. In the only available study on reforestation, J. P. Lanly showed in 1982 that over half of all virgin rain forest cleared turned not into wasteland but into second-generation forests. These new forests usually support less diverse ecosystems than the virgin forests, but they do provide a buffer of support in which some endangered species can find safe haven. Furthermore, nearly 6 percent of the rain forests—about 1.8 million hectares—are in protected parks, and more land is being added all the time. Much of this land has been carefully chosen to protect both rare species and overall biodiversity. Also, several million rain-forest hectares are inaccessible to people or impractical for human use. These areas are not likely to be disturbed soon.

But all this good news isn't what some conservationists want you to hear. Don't get me wrong: there are many good reasons to be opposed to cutting down the rain forests, and I am absolutely against any destruction of virgin woodlands. I have emphasized the positives simply because conservationists often downplay scientific uncertainties and manipulate the numbers to make things look far *worse* than they actually are.

The island model estimates of the number of rain-forest organisms vary greatly, running from 5 million to 100 million species. (By contrast, the total number of species which have been described by science, including microscopic organisms, is less than 1.4 million.) Clearly, the uncertainties in these calculations are huge.

Just how many species are becoming extinct each year is even more uncertain. Even the experts disagree on how much of the forests we are devouring annually. The numbers I've found so far from "official" sources range from 0.5 percent to 1.8 percent forest cover lost each year. Some environmentalists—such as E.O. Wilson of Harvard University—estimate that we are "easily" losing 100,000 species every year. Others use the above deforestation rates to argue that the number is between 6,250 and 22,500. But even these guesses are likely to be too high, because they assume a species area curve that we know *overestimates* rain-forest biodiversity. What's more, to obtain these frightening rates, one must apply the island model in the most naive way—with no account of forest regeneration, the pattern of clearing, animal migration, or a host of other critical details. While it is true that continued clearing must at some point drive species to extinction, we simply don't know at what point that is. In some countries, the damage is so great that it is hard to imagine that some extinctions are not taking place; but in others, where large sections of forest remain intact, the extinction rates may be closer to zero.

An Abundance of Bugs

Television ads targeted to elicit your financial support for one environmental group or another usually present the high extinction-rate numbers combined with footage of exotic animals and discussions of the cancer cures that might lie hidden in the plant species we are destroying. These ads create the impression that plants and higher animals are taking the brunt of the disaster, but that's all smoke and mirrors. Almost all of the rain forests' biodiversity lies within their bugs, and so even the island model predicts that the vast majority of species being extinguished are not plants that could cure cancer but *insects*. As unpopular as this may be to say, nature is so bug abundant (more than 600,000 species of beetles have already been described and there are likely to be many more out there) that losing a few percent of these is not likely to cause an ecological collapse. However, resorting to scare tactics in our attempts to save the rain forests could wind up being ecologically disastrous.

In his masterful book *The Mismeasure of Man*, Stephen Jay Gould documents just how badly biology has been misused by some scientists to prop up the prevailing prejudices of the day. Although Gould focuses on nineteenth-century biology, it would be folly to presume that no modern scientist perpetrates similar abuses.

Perhaps the field most abused these days is ecology, which sometimes seems to me more like a religious movement than a scientific discipline. Any "heretic" who dares stray from the most apocalyptic interpretations of the data risks the professional scorn of Mother Nature's self-appointed, doctoral-wielding apostolates. Several biologists who criticized the high species-extinction estimates have been snubbed at professional conferences and vilified in print. A.E. Lugo, project leader for the Institute of Tropical Forestry in Puerto Rico, reports that, after giving a speech at the Smithsonian Institution in which he criticized the high extinction rates usually quoted, a prominent conservationist verbally assaulted him in the cafeteria. Paul Ehrlich, the well-known Stanford biologist who is perhaps the new enviro-religion's foremost spokesperson, has even declared that we must all undergo a "quasi-religious transformation" aimed at creating a "revolution in attitudes toward other people, population growth, the purpose of human life, and the intrinsic value of organic diversity" if humanity is to survive.

This kind of rhetoric really frightens me. Why? Because the rain forests *are* mortally imperiled. More than half the virgin forests have already been cleared and many areas have already been devastated. Stopping the destruction will require us to be very careful about our facts. Alarmist statistics, dire predictions that don't come true, and petty self-righteousness risk the credibility of the whole cause of rain-forest conservation. Without credibility, there is no persuasion, and, without persuasion, there is absolutely no hope for us to save the forests.

Human Issues

We cannot save the rain forests without addressing the problems of their 100 million human inhabitants. It's pretty tough to sell hungry parents on the idea that the next acre of virgin forest they want to clear has a greater right to exist than their own families, and professional conservationists know this. That's why most programs to save the rain forests have the primary goal of improving agricultural productivity to reduce the need for further forest-clearing. Modern land-management techniques—the use of fertilizers to replenish the nutrients leached from the soil, tractors to break up the earth, careful crop rotation, and the judicious use of pesticides to maintain soil viability and reduce destruction from pests—could greatly reduce the need to clear virgin forests. Wastelands produced by logging can often be replanted with fast-growing native trees of high lumber value to provide loggers with a source of timber which can be more profitably harvested than virgin forests. Although only about 13 percent of the cultivated land is currently being managed to produce a sustainable yield, several comprehensive plans—like the United Nations' Tropical Forestry Action Plan—are now being implemented to slow the destruction. International strategies to reverse the rampant destruction of the world's rain forests are in place. We must now find the political will to implement them.

Thus, everything is not gloom and doom. The forces of social corrosion—poverty, governmental corruption, and foreign debt—are still hard at work conspiring to rid the world of our most biologically bountiful treasures, but there is plenty of reason for hope. What may be most dangerous and perhaps most difficult to overcome is our own tendency toward extremism. The winds of irrationality lash the public debate and make impossible the clear-headed discussion of the reasoned and practicable steps we must take to save the rain forests. We must choose data over dogma if the rain forests are to stand a chance.

"Though the older breeds hold genetic riches that one day may be priceless . . . they are endangered animals without an act of Congress to protect them."

The Extinction of Livestock Breeds Is a Serious Problem

Lisa Drew

Among domesticated animals, a breed is similar to a subspecies of a wild animal. Just as a certain type of squirrel or owl may be close to extinction while other members of the species are flourishing, so, too, may individual breeds of cows, horses, and other livestock be endangered. Lisa Drew, senior editor of *National Wildlife*, maintains that many endangered livestock breeds carry unique genes and traits that are often not found in newer breeds—traits that will be irrevocably lost should these breeds become extinct. Some organizations, such as the former American Minor Breeds Conservancy (now called the American Livestock Breeds Conservancy), are attempting to save the endangered breeds, Drew writes, but far more extensive efforts are necessary.

As you read, consider the following questions:
1. What advantages do older breeds have over modern super breeds, according to Drew?
2. What does Drew cite as major threats to endangered livestock breeds?

Lisa Drew, "In Search of the Barnyard Ark," *National Wildlife*, December/January 1991. Copyright 1991 by the National Wildlife Federation. Reprinted with permission.

Not far from where the Pilgrims came ashore to settle the New World, a dozen students take turns driving a team of Dutch Belted oxen. With spoken commands and a whiplike goad, they direct the yoked pair around a field. The overall effect on this sunny spring day is one of charm. The animals start, turn and stop in tandem, compliantly breaking in one handler after another. Their coloring, black interrupted by white bands encircling their middles, is echoed in the robes of one of the participants, a Benedictine nun. A student nervously yells, "Whoa!" near the beasts' sensitive ears, and the teacher says, "You don't have to be loud. I always say an ox's favorite command is 'Whoa.'"

Charm is important here. In another century, these animals might have been prized as hardy sources of meat and milk as well as brute strength. But the year is 1990, the place is Plimoth Plantation living history museum and the occasion is an American Minor Breeds Conservancy (AMBC) conference. These students, mostly hobbyists, small farmers and historians, know that almost half the breeds of livestock in this country are near the brink of extinction—and that almost no one else is doing anything about saving them.

Super Breeds

The modern farm has little use for animals like Dutch Belted cattle, as capable of pulling a plow as giving milk. Instead, it has developed single-purpose super breeds, like Holstein cattle for milk and Leghorn chickens for eggs. Though the older breeds hold genetic riches that one day may be priceless—for use in low-input farming, adaptation to climate change, even conservation and maintenance of wildlife habitat—they are endangered animals without an act of Congress to protect them. Many have survived only because they happen to have enchanted their keepers.

"We spend the least amount of money for our food—for probably the highest quality of food—anywhere in the world," says AMBC executive director Don Bixby. "How can you complain about that? But there are expenses that haven't been put on the ledger sheet, and genetic conservation is one of them."

Livestock conservation is not this haphazard everywhere in the world. It is most organized in Great Britain, which has a model nonprofit program called the Rare Breeds Survival Trust, and in the Eastern Bloc countries, which sponsor state-run programs. Some countries—most notably Hungary and Brazil—boast government farms for rare livestock. And the United Nations Food and Agriculture Organization has solicited livestock surveys from around the world, starting in developing countries. The United States now accounts for the biggest gap in the data.

The U.S. Department of Agriculture has been doing genetic research on some common and rare breeds of livestock for more than 30 years, and in 1991 it released a national plan for animal gene conservation, including its first census. But as for actually saving breeds at risk, "We're just talking about getting all this pulled together," says Roger Gerrits of the USDA Agricultural Research Service. Eventually, the agency may store frozen semen and embryos of rare breeds. "After all is said and done," he adds, "a lot more has been said than has been done."

The closest thing in the United States to a livestock Noah's ark is the American Minor Breeds Conservancy, based in Pittsboro, North Carolina. During the 1980s, the nonprofit organization helped give a sense of mission to owners of rare breeds, growing from 150 members to 3,000. "I'm raising American heritage farm animals," says AMBC member Margaret Marsh of Princeton, Massachusetts, whose 2 acres of pasture hold rare breeds of cattle, pigs and sheep between colonial stone walls. "What we're doing now may help farmers 50 years from now."

The AMBC has started some limited semen banking for certain breeds, but it has no funds for actually keeping animals; that is up to individuals and luck. And only since the organization conducted a 1985 census on a shoestring budget have members had any idea what breeds survive in this country and where they are.

During the weekend conference at Plimoth Plantation, 150 AMBC members ponder such issues as the future of draft animals and the importance of feral livestock, animals that have reverted to the wild and been toughened by the rigors of natural selection. Between meetings, participants show each other photographs of their animals, pulling snapshots from wallets and hauling out whole scrapbooks. In a very real sense, this is when they do some of their most important work: Survival for many dwindling breeds often means finding unrelated mates to forestall inbreeding.

Living Links

Losing the old livestock could be a tragedy on a huge scale. The animals in the photographs—four-horned sheep, shaggy-haired cattle, draft horses, reddish pigs—are largely irreplaceable. Some are the closest living links to the wild animals from which domestic animals were first tamed thousands of years ago.

Some breeds are traceable to their wild roots: The ox ancestor of most of the world's cattle went extinct in the 1600s. Others are mysteries: Scientists can only guess at the ancestry of domestic horses. One theory holds that domesticity actually saved horses as forests and human hunters encroached on the grass-eaters at the end of the Ice Age. Only one truly wild horse still lives, the Przewalski's, which roamed Asia 60 million years ago. Ironically, the 1,000 surviving Przewalskis exist only in semi-captivity, while once-tame feral horses (like mustangs on our plains) run wild in parts of the world.

Dwindling Numbers

Legions of domesticated animals across the globe—pigs and cattle, sheep and goats, poultry, horses and ponies—are gravely diminished in number or on the verge of vanishing altogether. . . .

In Western Europe, 230 native breeds of cattle existed at the turn of the century: 70 of these had become extinct by 1988, and 53 were endangered.

Kathleen Burke, *Smithsonian*, September 1994.

Many of the old breeds are a form of living history. The Crusaders used Percheron draft horses on the way to the Holy Land; they later became our most popular draft horses. In 1985 there were about 1,500 registered here, up from 85 in 1954. The sunburn-resistant Tamworth pig, one of the oldest surviving pigs, is similar to the domestic swine of medieval times, which in turn descended from ancient wild boars. About 2,000 were registered in this country in 1985. The gentle Guinea hog, once a common homestead pig in the South—and now almost extinct—came to this continent with the slave trade. Almost all livestock on this continent originated with animals that accompanied Spanish and European settlers or slave traders. Some of our minor breeds still exist in Europe, but generally are rare there too.

There is far more at stake than cultural history or a few unique traits. Until the farm technology of the past few decades, livestock survival had a great deal to do with natural selection—despite selection by breeders for productivity or more frivolous traits (like the markings of Dutch Belted cattle). Older breeds tend to forage well for their own food, resist disease and mother their young without help. They also were once key links on farms with many interlocking parts, consuming byproducts like whey and cider pulp, fertilizing fields, even controlling pests.

Over thousands of years, some farm animals have become integral parts of ecosystems. Some are named for the regions where they evolved, such as British sheep called Cotswold, Dorset or Oxford. Their counterparts among wild animals are subspecies, populations exquisitely adapted to specific habitats.

After an important Indian waterfowl reserve banned grazing in 1982, bird populations plummeted; the Bombay Natural History Society later blamed in part the withdrawal of water buffalo that had controlled aquatic grasses. Wildlife trusts in Britain advertise land available for grazing by rare breeds to help maintain habitat. In the United States, farmers have experimented with using Scotch Highland cattle in place of herbicides to control unwanted plants like poison oak and blackberry bushes. The cattle are also known for their lean meat; their long coats serve in place of fat to warm them.

Factory Farm Animals

The habitat of animals on the modern farm, in contrast, is often a sort of McDonald's environment, the same place no matter where it happens to be on the map. Animals adapt to antibiotics, assisted birthing, heated barns and high-energy foods rather than to weather or local plants. "You can go anywhere in the world now, and an intensive poultry farm will look identical in Iowa, Cotswold Hills and Addis Ababa," says Elizabeth Henson of Britain's Rare Breeds Survival Trust. "There are the same cages, same heating, same formula food and exactly the same birds. That is totally different to what it would have been like even 50 or 60 years ago, when every village in the world would have had slightly different birds."

Such farms generate enormous amounts of food; the worry is about what is being lost in the process. "Today's animals work extremely well for today's environments," says geneticist Phillip Sponenberg of Virginia-Maryland Regional College of Veterinary Medicine. But, he says, if we create a population of uniform animals, "We may not be able to redirect it."

One example may be the commercial turkey, one of the poultry industry's great success stories, which has such a large breast that it can't reach the opposite sex to mate. "There are very few turkeys, except maybe in somebody's farmyard, that can mate naturally," says George Farnsworth, chief geneticist of Nicholas Turkey Breeding Farms in California. Although turkeys are a marvel of modern breeding, with huge breasts of lean white meat and white feathers that leave the skin unmarred by color, their inability to mate is part of a general trend that concerns AMBC. "They're being modified out of any form of adequate biological fitness," says poultry geneticist Roy Crawford, "to meet a commercial demand for right now."

That notion is absurd to people like Farnsworth, who calculates intricate genetic formulas to increase the birds' vigor, size and efficiency. The company keeps more than two dozen strains as genetic libraries (and corporate secrets) to draw on in the future. But even among those breeds, only one or two can mate naturally. And, says Farnsworth, "If some strains don't seem to have any potential in a real way, we get rid of them."

That mentality, certainly common for a corporate world of what Farnsworth calls "cutthroat competition," is precisely what worries livestock conservationists. Says Henson of the Rare Breeds Survival Trust, "We owe it to the next generation to hand them the same package of genes that was handed to us. It is very arrogant to say that people 100 years from now will want the same turkeys we eat and will want to continue to do artificial insemination."

If so, whose arrogance is it? If the case for keeping rare breeds is compelling, then can we blame someone for their status? "That is hard to do," says AMBC's Bixby. "Our society, and the short view of the profit motive, are to blame. Genetic conservation is one of those things we didn't know we needed to worry about."

Sustainability

Raising rare breeds is, all agree, no way to get rich. But who is to say that farmers of the future won't once again value abilities beyond food and fiber production? "These breeds evolved in an era of sustainability," says Bixby. "If we are to reachieve that sustainability, they will have a commercial niche." On the largely self-sufficient farm of the Abbey Regina Laudis, a Benedictine order in Bethlehem, Connecticut, Sister Telchilde Hinckley purchases no commercial feed for the Tamworth pigs. Instead, she gives them kitchen scraps, skim milk and whey from the dairy and bakery waste. "In very severe winters," she has found, "Tamworths have outwintered Yorkshires by a mile."

In 1989, Rodale Research Center in Pennsylvania tried controlling an orchard's weevil pest with commercial chickens. "They were pretty pathetic," says Orchard Project Manager Sarah Wolfgang. "The first few weeks they were here, we had to literally pen them out of the shelter during the day."

The problem was no surprise to AMBC when Rodale contacted the organization for help. "The commercial chickens didn't understand the issue," says Bixby. In 1990, Wolfgang tried Dominique chickens—known as intelligent and self-sufficient. Though it is too early to tell whether the animals will control the pests, "These chickens are certainly better adapted," says Wolfgang. "They're excellent foragers, and they range across three connected plots instead of huddling in a group. They seem quite content in the environment."

That connection with the environment is key in developing countries, where well-meaning aid agencies are bringing in productive livestock to replace local breeds—without enough attention, say critics, to whether regions can support the coddled environments the animals need to survive. "The indigenous breeds," says poultry geneticist Crawford, "are falling like ten pins." There are no figures to support such charges, but there is plenty of anecdotal evidence. Just one example: In Africa's Chad Valley, commercial cattle largely replaced small local stock in the early 1970s. Then the region's usual seven-year drought hit, and, "Surprise, surprise," says Henson, "the imported stock died." Farmers rebuilt their herds with a few remaining local cattle. . . .

Threats to Minor Breeds

Livestock conservationists worry that we may not have much time left in this country to save our equivalents of Chad Valley cattle, as many of our rare breeds are in the hands of aging farmers. "It doesn't take a lot of thought to realize those herds are going to be gone if we don't do something about it," says geneticist Sponenberg. One of those farmers is 69-year-old James F. Holt of Rochelle, Georgia, whose herd of Pineywoods cattle (also known as Florida Cracker cattle) has been in his family for 200 years. Fewer than 500 of the animals, descended from Spanish cattle brought here in the 1500s, survive in the whole country.

The cattle "practically feed themselves," says Holt, and calve regularly until the age of 25. Sponenberg calls them "indestructible." The last time Holt needed a veterinarian was five years ago, and that was for a crossbred animal. Other cattle may seem more productive, he says, "but here's the catch: In those Pineywoods cattle, you can feed five for the expense of feeding three of the others."

Holt's simple economics aside, herds like his could easily fall victim to the market—or worse. "There are dreadful consequences to a rare breed when even one key herd is forced to liquidate due to financial distress," says Canadian pig farmer Douglas Law of Moffat, Ontario. In the United States, Dutch Belted cattle suffered a blow in the 1986 dairy buy-out program. To relieve the glutted dairy market, the federal government offered cash for cows. Many older farmers retired, including a Florida farmer who sold about 1,000 Dutch Belts for slaughter. By 1987, fewer than 1,000 purebred Dutch Belts survived in this country.

In an even sadder chapter of U.S. history, the federal government almost killed off the Navajo Churro sheep, a particularly hardy breed with distinctive two-textured wool (wool that helped make Navajo textiles famous). The sheep became an integral part of Navajo culture after the Indians acquired them from Spanish settlers early in the 17th century. After the U.S. Army declared war on the Navajo nation in 1863, fighting decimated livestock populations. Then, between 1930 and 1950, the federal Navajo Livestock Reduction Program—an effort to reduce overgrazing on limited reservation land—indiscriminately slaughtered both good and poor Churro stock. Probably fewer than 1,000 still exist [as of 1991], up from about 500 in 1985.

The only farm animal the U.S. government has actually saved is the Texas Longhorn. In 1927, 20 found a home on a wildlife sanctuary after Congress cited the animal's historical significance. Today, the demand for leaner beef has again made Longhorn genes commercially viable.

Although a one-time save by the government, the Longhorn case has much in common with other efforts to rescue breeds in trouble—efforts called "fire brigade stuff" by Elizabeth Henson. "In a way we're in the same position in livestock conservation as the World Wildlife Fund was when it first looked at individual species," she says. "It's hard to step back far enough to see the whole picture.". . .

Saving Genetic Diversity

Says AMBC's Bixby, "Conservation is sometimes practiced not because of any idea that one species or breed is more important than another, but just because it's threatened. The point is, we need to save genetic diversity."

Even fainting goats, which fall over with rigidly contracted muscles when startled? (The condition must set predators to licking their chops.) Yes, says Bixby, though even he agrees that the animal's peculiar condition "has no agricultural value." The goats are still sources of meat and milk, and they do have a place in medical research, as people can also be affected by the hereditary condition, called myotonia. Bixby insists that they are also valuable just because they are "different in a measurable, discernible way, and they may have some other characteristics we don't know about that will be valuable."

Philosophizing aside, for the time being the key to the goats' survival is simply that they charm us. And that, whimsical as it seems, continues to be critical not only for fainting goats but for the survival of this country's whole barnyard ark.

VIEWPOINT 6

"The primary sources for all of our principal food crops are being systematically destroyed."

The Extinction of Wild Plants Threatens Agriculture

Al Gore

Al Gore, a former senator from Tennessee, was elected U.S. vice president in 1992. In the following viewpoint, excerpted from his book *Earth in the Balance*, Gore contends that wild varieties and relatives of the world's food crops are in grave danger of extinction due to human encroachment on their habitats. These wild grains, fruits, and vegetables are essential to humankind, Gore maintains, because they contribute to genetic diversity and carry useful traits that are not found in hybrid crops created through biotechnology. Loss of these wild plants would increase the vulnerability of the world's food supply to disease, drought, and pests, Gore argues.

As you read, consider the following questions:

1. In the author's opinion, what are the shortcomings of new crop varieties developed with biotechnology?
2. How does genetic erosion threaten the world's food supply, in Gore's view?
3. How have economic problems in impoverished countries contributed to genetic erosion, according to Gore?

Excerpted from *Earth in the Balance* by Al Gore. Copyright ©1992 by Senator Al Gore. Reprinted by permission of Houghton Mifflin Company. All rights reserved.

Every seed (and seedling) carries what is known as the germplasm; it contains not only genes but all of the special qualities that control inheritance, define the way genes work, and fix the patterns by which they combine and express their characteristics—in the words of expert Steve Witt, "the stuff of life." But the future health of the food supply depends on a wide variety of this irreplaceable stuff, and we now risk destroying the germplasm that is essential to the continued viability of our crops. Crucial to any food supply is the genetic resistance of crops to massive destruction from blights, pests, and changing climate. Maintaining genetic resistance requires the constant introduction of new strains of germplasm, many of which are found only in a few wild refuges around the world. These fragile places serve as the nurseries and reservoirs of genetic robustness, vitality, and resilience, but all of them are now in serious jeopardy. Indeed, the primary sources for all of our principal food crops are being systematically destroyed. This danger is only just now being understood by agronomists; one who does is Te-Tzu Chang, the head of the International Storage Center for Rice Genes in the Philippines, where 86,000 varieties of rice are kept. As he told *National Geographic*, "What people call progress—hydroelectric dams, roads, logging, colonization, modern agriculture—is putting us on a food-security tightrope. We are losing wild stands of rice and old domesticated crops everywhere."

Biotechnology is, to be sure, creating new crop varieties with impressive characteristics, such as uniformity, high yield, and even natural resistance to crop diseases and pests. But we have been blind to the harsh truth that the new crop varieties we create in our laboratories quickly become vulnerable to their rapidly evolving natural enemies, sometimes after only a few growing seasons. Although their genetic resistance is reinforced with new genes that are spliced into the commercial varieties every few years, many of the genes available for replenishing the vitality of food crops exist only in the wild. . . .

Vavilovian Centers

Finding these wild strains is often not a simple matter. Plant geneticists must literally return to the specific place on earth where the endangered crop makes its genetic "home" and search through the countryside—sometimes on hands and knees—for a wild relative. These genetic homelands are also known as centers of genetic diversity, or Vavilovian centers, in honor of Nikolai Ivanovich Vavilov, the Russian geneticist who discovered and described them. As it happens, there are only twelve such centers in the world, each the ancestral home of a dozen or so of the most important plants to modern agriculture. The total number of important crops is remarkably small: virtually all of the world's food crops and feed grains come from only about 130 plant species, the vast majority of which were first cultivated in the Stone Age.

Most centers of diversity are found, as Vavilov said, "in the strip between twenty degrees and forty-five degrees north latitude, near the higher mountain ranges, the Himalayas, the Hindu Kush, those of the Near East, the Balkans, and the Apennines. In the Old World, this strip follows the latitudes, while in the New World, it runs longitudinally, in both cases conforming to the general direction of the great mountain ranges." The ancestral home of wheat, for example, is the mountainous terrain of northern Iraq, southern Turkey, and eastern Syria, squarely within the strip described by Vavilov. Many strains of wheat grow naturally here, but this variety is not reflected in domesticated wheat. Indeed, less than 10 percent of the genetic variety in wheat is found in the plants currently grown as crops. According to the biologist Norman Myers, another 30 percent of the genetic diversity in wheat can be found in the various seedbanks around the world. But almost two thirds of all wheat strains are found only in the wild, most of them still in the original Vavilovian center.

Losing Genetic Diversity

The widespread introduction of a handful of high-yielding, or "Green Revolution," crop varieties has boosted overall food production over the past several decades but eliminated many traditional strains that were well adapted to local ecosystems and could have been used to develop higher-yielding, locally appropriate crops. If current trends continue, three quarters of India's rice fields may be sown in only 10 varieties by 2005. In Indonesia, 1,500 local varieties of rice have disappeared [since 1977], and nearly three fourths of the rice planted today descends from a single maternal plant. Similarly, in the U.S., just six varieties of corn account for 71 percent of the corn fields, and nine varieties of wheat occupy about 50 percent of the wheat land.

John C. Ryan, *Life Support: Conserving Biological Diversity*, April 1992.

The center of diversity for coffee is in the Ethiopian highlands. But coffee is now grown in many areas of the world—the Andes region of Colombia and Brazil is one—and every once in a while, when a new pest or blight cannot be met with genetic resistance from readily available seeds, coffee growers must return to the Ethiopian highlands in search of wild relatives that can combat the new threat. A few years ago, this reliance on coffee's genetic homeland took an ironic twist. As Brazil was coming under international criticism for its tolerance of widespread destruction of the Amazon rain forest, a small group of Brazilians went to Addis Ababa to express their concern about the progressive deforestation of areas in Ethiopia vitally important to the future viability of the coffee crop.

In the case of corn, elevated portions of Mexico and Central America are home, while the potato is native to specific areas of the Andes in Peru and Chile. For centuries, even millennia, these remote centers of diversity were safe. Vavilov speculated that the Stone Age crops on which we so utterly depend today were able to survive in these mountainous regions because of their great diversity of soil types, topography, and climate. Furthermore, the inaccessibility of the mountains and the isolation of the valleys between them provided relatively good protection from the disruptions of civilization and commerce.

Vanishing Plants

Unfortunately, our global civilization has now acquired such enormous power and reach, and the demands of a burgeoning population for land, firewood, and resources of every kind and description are now so ravenous, that communities are encroaching rapidly on every single one of the twelve Vavilovian centers of diversity—even the most remote. For example, in Mesopotamia, the homeland of wheat, virtually the only areas where wild relatives of wheat can now be found are in graveyards and castle ruins. They survive because civilization, which often shows little regard for nature, at least sets aside tiny plots to commemorate its own past. But that is protection by accident, and too often these days we are relying on luck, not careful planning.

In an episode reported by Norman Myers, virtually the entire rice crop in southern and eastern Asia was threatened in the late 1970s by a disease called grassy stunt virus, which was spread by brown hopper insects. The threat to the food supply of hundreds of millions of people was so potent that scientists at the International Rice Research Institute in the Philippines frantically searched through 47,000 varieties in gene banks throughout the world for a gene to resist the virus. Finally, they found it in a single wild species from a valley in India. But this plant wasn't on sacred ground, and soon afterward, the valley was flooded by a new hydroelectric project. What if the same search had taken place today?

Recent history abounds with situations that show just how severe the strategic threat to our modern food supply has become. In 1970, the United States suddenly suffered devastating losses in its corn crop when the southern corn leaf blight took advantage of a trait that had been uniformly bred into virtually all of the corn crop in order to simplify the genetic manipulation itself. In 1977, scientists searching in Ecuador found a wild relative of avocado that was resistant to blight, a genetic trait of tremendous value to avocado growers in California. But the good news came with bad news: this strain of avocado was growing on only twelve trees in a tiny patch of forest, one of the last remnants of a large lowland forest that had been cut down to accommodate the needs of a growing Ecuadorian population. . . .

The short-term threat is, of course, not the extinction of important food crops, at least not as extinction is commonly understood. (Extinction is more a process than an event.) The way a plant or animal avoids extinction is by retaining enough genetic variety to adapt successfully to changes in its environment. If its range of genetic diversity is narrowed, then its vulnerability is correspondingly increased, sometimes to the point that a threshold is crossed and the complete disappearance of the species becomes inevitable. In every case, long before the last representative of an endangered species succumbs to its fate, the species itself becomes functionally extinct. The steady loss of genetic diversity in a species is called genetic erosion, and an astounding number of important food crops now suffer from it at a high rate. Among those listed by the United Nations International Board for Plant Genetic Resources as most at risk are the apple, avocado, barley, cabbage, cassava, chickpea, cocoa, coconut, coffee, eggplant, lentil, maize, mango, cantaloupe, okra, onion, pear, pepper, radish, rice, sorghum, soybean, spinach, squash, sugarbeet, sugarcane, sweet potato, tomato, wheat, and yam.

Genetic Diversity

During most of the history of agriculture, genetic diversity has been found not only among wild relatives of food crops but also among so-called landraces (also called primitive cultivars). These are plants genetically related to the food crops used in the global agricultural system that have been developed in more primitive agricultural systems. Neither as wild as their uncultivated relatives in the mountain valleys nor quite as refined as their modern hybrid cousins, they nevertheless contain a much wider range of genetic diversity than do advanced breeding lines. Unfortunately, many landraces are now also endangered because of the spread of modern, higher-yielding varieties. An international conference in 1990 in Madras, India, sponsored by the Keystone Center, concluded that "it is an unfortunate reality that many nations have knowingly or unknowingly lost their traditional landraces due to the spread of high-yielding varieties, thereby increasing genetic homogeneity." In the United States, for example, of all the varieties of vegetables listed by the Department of Agriculture in 1900, no more than 3 percent now remain, according to one estimate.

The United States, however, has only one center of diversity (the upper Midwest, where the blueberry, cranberry, Jerusalem artichoke, pecan, and sunflower originated). Virtually all of the other centers lie in Third World countries, surrounded by exploding populations searching for firewood, food, and land—even heretofore remote land—on which to live. In order to earn hard currency from the sale of exports and thereby finance their enormous debts to the industrial nations, these impoverished countries are taking land formerly used for subsistence agriculture—many of them with genetically rich landraces—and using them instead for growing monocultured hybrid crop varieties for sale overseas. (This pattern has precedents. During the Great Potato Famine, for example, Ireland was growing a lot of wheat, but almost all of it had to be exported to pay its debts to England.) To be sure, these same new "miracle crops" also provide higher yields for domestic markets and have temporarily conquered hunger in a few of the Third World nations. But the much-heralded Green Revolution has, in most countries, failed to address fundamental economic problems, such as those caused by inequitable land ownership patterns, which often allow a wealthy elite to control a huge percentage of the productive land. And some of the ballyhooed development programs organized and funded by international financial institutions become part of the problem as well: in too many cases they turn out to be wildly inappropriate for the culture or ecology of the region in which they are placed. Moreover, the higher yields made possible by genetically altered crop strains often cannot be sustained over time, as the pests and blights catch up to them and as overirrigation and overfertilizing take their toll on soil productivity. . . .

The Value of Wild Genes

It is virtually impossible to calculate the value of maintaining the rich diversity of genetic resources on earth. And indeed, their value cannot be measured by money alone. But where food crops are concerned, we at least have some yardsticks with which to approximate the value of genes that are now endangered. The California Agricultural Lands Project (CALP) recently reported that the Department of Agriculture searched through all 6,500 known varieties of barley and finally located a single Ethiopian barley plant that now protects the entire $160 million California barley crop from yellow dwarf virus. Similar wild genes have contributed to increases in crop yields—more than 300 percent in many crops—in the last few decades. Among the many examples of the value of wild genes found by the CALP are "a seemingly useless wild wheat plant from Turkey [that] passed disease resistant genes to commercial wheat varieties worth $50 million annually to the United States alone; and, a wild hop plant [that] gave 'better bitterness' to English beer and in 1981 brought $15 million to the British brewing industry."

The value of genetic diversity has been noted, of course, by those who invest in global agriculture as well as by plant geneticists. For that reason, there is now another source of diversity besides the wild relatives and landraces: gene banks, an amazing variety of them. Some are managed by governments, some by private seed companies and multinational corporations, some by universities, and a surprising number by individuals, many of whom are merely dedicated hobbyists. The current system is in scandalous condition, with insufficient government attention and money, little coordination between different repositories, grossly inadequate protection and maintenance of national collections, and a missing sense of urgency where such a precious resource is concerned—especially with respect to the many vegetables and grains that presently play a smaller role in world agricultural trade and are thus at even greater risk. . . .

Nature Creates Genes

But the single most serious strategic threat to the global food system is the threat of genetic erosion: the loss of germplasm and the increased vulnerability of food crops to their natural enemies. Ironically, this loss of genetic resilience and flexibility is occurring at precisely the moment when those who believe that we can adapt to global warming are also arguing that we can genetically engineer new plants that will thrive in the unpredictable conditions we are creating. But scientists have never created a new gene. They simply recombine the genes they find in nature, and it is this supply of genes that is now so endangered.

Our inability to provide adequate protection for the world's food supply is, in my opinion, simply another manifestation of the same philosophical error that has led to the global environmental crisis as a whole: we have assumed that our lives need have no real connection to the natural world, that our minds are separate from our bodies, and that as disembodied intellects we can manipulate the world in any way we choose. Precisely because we feel no connection to the physical world, we trivialize the consequences of our actions. And because this linkage seems abstract, we are slow to understand what it means to destroy those parts of the environment that are crucial to our survival. We are, in effect, bulldozing the Gardens of Eden.

Periodical Bibliography

The following articles have been selected to supplement the diverse views presented in this chapter.

Y. Baskin	"Ecologists Dare to Ask: How Much Does Diversity Matter?" *Science*, April 8, 1994.
Elissa Blum	"Making Biodiversity Conservation Profitable," *Environment*, May 1993.
Paul R. Ehrlich and Gretchen C. Daily	"Population Extinction and Saving Biodiversity," *Ambio*, May 1993. Available from 810 E. 10th St., Lawrence, KS 66044.
A.M. Gillis	"Keeping Traditions on the Menu," *BioScience*, July/August 1993.
Indur M. Goklany and Merritt W. Sprague	"Limits on Technology Would Hurt Biodiversity," *USA Today*, November 1992.
Oona Hathaway	"Whither Biodiversity? The Global Debate over Biological Variety Continues," *Harvard International Review*, Winter 1992-93. Available from PO Box 401, Cambridge, MA 02238.
Tamar F. Hurwitz	"Species Extinction in the Rainforest," *The Animals' Voice*, July/August/September 1994. Available from PO Box 16955, North Hollywood, CA 91615.
Jonathan W. King	"Breeding Uniformity," *The Amicus Journal*, Spring 1993.
Ariel E. Lugo, John A. Parrotta, and Sandra Brown	"Loss in Species Caused by Tropical Deforestation and Their Recovery Through Management," *Ambio*, May 1993.
Julian L. Simon and Aaron Wildavsky	"Facts, Not Species, Are Periled," *The New York Times*, May 13, 1993.
Nigel J.H. Smith, J.T. Williams, and Donald L. Plucknett	"Conserving the Tropical Cornucopia," *Environment*, July/August 1991.
William K. Stevens	"Species Loss: Crisis or False Alarm?" *The New York Times*, August 20, 1991.

CHAPTER 2

Can Endangered Species Be Preserved?

Endangered Species

Chapter Preface

At the Seney National Wildlife Refuge in Michigan, a biologist, covered from head to toe in a brown sack, puts on a hand puppet that resembles the head and neck of a sandhill crane. Wearing this disguise, the biologist interacts with young, parentless sandhill cranes, teaching the captive birds techniques they will need to survive in the wild. If this attempt at surrogate parenting succeeds, researchers plan to use the puppet rearing method with the endangered whooping crane—a bird that has not responded well to earlier, more conventional recovery measures.

Playing "Mom" is just one of the many extremes that scientists and ecologists go to in order to save endangered species. Artificial insemination and embryo transplants have been implemented to prevent inbreeding, which can weaken a species' chances for survival by increasing the likelihood of inheriting genetic defects. Federal and state wildlife agencies sometimes poison predators that threaten the survival of an endangered species. The Endangered Species Act protects endangered species from being hunted, collected, injured, or killed on federal and public lands. Concerned that an insufficient number of species were being added to the endangered species list during the 1980s, a coalition of environmental groups even sued the U.S. Department of the Interior. The lawsuit was settled out of court in 1992 with an agreement that the U.S. government would add 382 species to the endangered species list by September 30, 1996.

However, these methods have not always produced the desired result of aiding endangered species. Despite the use of sophisticated reproductive technology, many endangered animals have had far less success breeding in captivity than in the wild. Conservationists have been stymied in cases where one protected species jeopardizes another, as when California sea lions began to devour large quantities of steelhead trout. Moreover, as Timothy Noah reports in the *Wall Street Journal*, the 1992 legal settlement resulted in "heightened regulatory activity [that] added to a political backlash against . . . the Endangered Species Act." Many politicians question whether the provisions of the act have been effective in helping species. In his book *No Turning Back: Dismantling the Fantasies of Environmental Thinking*, Wallace Kaufman argues that "the Endangered Species Act has not saved any species" and calls for less government involvement in the preservation of endangered species.

In their efforts to save endangered species, conservationists must determine which methods are most effective. The authors in the following chapter debate the strengths and weaknesses of several measures used to save endangered species.

1 VIEWPOINT

> "The Endangered Species Act [has] failed to recover or save a species in two decades."

The Endangered Species Act Is a Failure

Robert E. Gordon

The 1973 Endangered Species Act, which enables the U.S. government to designate and protect endangered and threatened species, comes up for reauthorization every four years. The act was due for reauthorization in 1992, but congressional debate was postponed until 1995 and was pending at the time this viewpoint was written. Many people believe that the act has not been effective and think it should be amended or discarded. Robert E. Gordon, executive director of the National Wilderness Institute, argues in the following viewpoint that the Endangered Species Act has cost the United States millions of dollars without recovering a single species. Furthermore, Gordon contends, private interest groups and property owners could do a much better job of saving species were they not penalized by the law.

As you read, consider the following questions:

1. In Gordon's opinion, what are the actual reasons for the recovery of the six species that were removed from the endangered species list?
2. How do nonendangered species end up on the endangered species list, according to the author?
3. On what basis does Gordon object to the costs of implementing the Endangered Species Act?

Robert E. Gordon, "Help Landowners Save Endangered Species," *Insight*, May 30, 1994. Reprinted with permission from *Insight*. Copyright 1994 *Insight*. All rights reserved.

Politically, we live in an age of platitudes. Nowhere is this more true than in the case of the environment—save the Earth. Even the names of major federal regulations have a similar ring— Clean Water Act, Clean Air Act, Safe Drinking Water Act. The Endangered Species Act is a perfect example. The cry is that we must save endangered species, and on face value who could question that? But as is so often the case in Washington, the distance between reality and the laws written within the Beltway can be pretty far.

If you examine the three words of environmental lingo—save, endangered and species—and the current federal programs designed to protect our planet, some rather disturbing facts will surface. Consider the programs involving the word save. The act has been an utter failure. After more than two decades of implementation, not a single plant or animal has recovered as a result of the act and then been removed from federal protection. The word endangered hasn't been used well either. Many animals and plants covered by this federal do-good venture are not endangered at all, a fact which the government grudgingly admits on occasion. And the use of the word species has nothing to do with science. It seems many members of Congress learned little in high school biology. Many of the plant and animal species we are supposed to be saving are not really species at all. As if these problems are not enough, the costs of this botched program are beginning to loom large even by Washington standards. The costs of this multibillion-dollar federal boondoggle are even more alarming considering the loss of private-property rights. Private stewardship of the land may be one of the things that could actually benefit truly endangered species.

Undercounted Species

Under the Endangered Species Act, the U.S. Fish and Wildlife Service, or FWS, and the National Marine Fisheries Service, or NMFS, are supposed to identify plants and animals headed toward extinction, add them to an official list, fix them up and then remove them from the list when they are "recovered." Sounds fairly reasonable. But, in application, it is a bit different. After two decades of implementation, the government only claims six plants and animals as "recovered." Three of these are birds—the Palau dove, Palau owl and Palau flycatcher—which live on a small island trust territory about 400 miles east of the Philippines. While the wildlife service calls them recoveries, a General Accounting Office report states that "although officially designated as recovered, the three Palau species owe their 'recovery' more to the discovery of additional birds than to successful recovery efforts." In short, they never were endangered in the first place, just undercounted.

Similarly, a former FWS director revealed during a Senate hearing that a plant known as the Rydberg milk vetch was called a recovery because "further surveys turned up sufficient healthy populations"—again, a mistake. Similar to other recovered species, the American alligator should not have been added to the endangered species list. Even the National Wildlife Federation admitted in its magazine that the "familiar and gratifying" recovery story of the alligator was "mostly wrong." But they have been hypocritically promoting the alligator as a success story now that the act is up for reauthorization.

The Eastern population of brown pelicans is one of the few species (actually a population of a species) which may have merited addition to the list and then removal because its numbers increased. The improvement, however, is not attributable to the Endangered Species Act. Most people credit a concurrent ban on the pesticide DDT. The official list of the species recovered and removed from the list ends here.

A few other species are often cited as successes, including the Arctic peregrine falcon, the bald eagle, and the gray whale. However, these claims are also disingenuous. In the case of the falcon, the wildlife service's new director sent out a press release attributing the growing population to the unrelated DDT ban. The ban is also considered by many as one of the primary reasons the bald eagle's numbers have increased. While the gray whale is currently at an all-time population high in the Pacific, its population has grown since 1890 (83 years before the act existed). Certainly, the success of the whale is not due to the act.

Two Decades of Failure

Not only has the Endangered Species Act failed to recover or save a species in two decades, but also the odds seem slim that this record is going to improve. The wildlife service cannot even demonstrate that a species' condition generally improves when it is added to the federal list. Some proponents of the current act are claiming that it just takes time and that the status of 40 percent of federally listed species are "improving" or at least "stabilized." These figures, however, are only estimates by FWS officials with no solid numbers to back them up.

And many of the 800-plus plants and animals on the federal list do not need saving—they are not endangered. Not only did the government wrongly regulate the alligator, the three birds from Palau and the Rydberg milk vetch, but also many other listed plants and animals. Federal regulatory agencies make a determination about whether to add a plant or animal to the federal list on the basis of the "best available data." The law does not require accurate, verifiable, reliable or conclusive scientific evidence of species endangerment.

One of three grounds for removal from the list of endangered species is data error. The fact that this category even exists demonstrates that solid scientific information is not required for a listing. Six of the 18 species removed from the list were officially recognized as mistakes. At an estimated $60,000 per listing and $37,000 per cross-off—not to mention incidental costs—these errors add up.

Mistakes

Regarding one specific data error case, the Federal Register states, "As a result of the Indian flapshell turtle's inclusion on Appendix I of CITES [a United Nations endangered species list] the Service subsequently listed the species as endangered." After listing (rather than before), a review was conducted "to see if supporting evidence justified its current endangered status. No such supporting data could be found." FWS then contacted turtle experts such as E.O. Moll, who stated that it was "seemingly the most common and widespread turtle in all of India.... How it ever made Appendix I is a big mystery."

Top 10 List of Endangered Species by Total Spending, 1989–1991.

Bald Eagle	$31,300,000
Northern Spotted Owl	$26,400,000
Florida Scrub Jay	$19,900,000
West Indian Manatee	$17,300,000
Red-Cockaded Woodpecker	$15,100,000
Florida Panther	$13,600,000
Grizzly (or Brown) Bear	$12,600,000
Least Bell's Vireo	$12,500,000
American Peregrine Falcon	$11,600,000
Whooping Crane	$10,800,000

Andrew Metrick and Martin Weitzman, Harvard University, 1994.

The tumamoc globeberry was crossed off the list by FWS on June 18, 1993. After including this plant on the endangered species list for seven years, FWS determined that "surveys have shown tumamoc to be more common and much more evenly distributed across its range than previously believed." Although never really endangered, during its seven years on the list, this plant soaked up more than $1.4 million from seven government agencies and was the Fish and Wildlife Service's basis to obstruct construction of the Tucson Aqueduct.

Erroneous listings include the Mexican duck, the pine barrens tree frog, and McKittrick's pennyroyal—a plant of the mint family. Other erroneous listings not even considered for removal include the northern spotted owl, of which there are now four to six times as many known to exist than was previously believed. (Were the government to admit this mistake, public support for the Endangered Species Act and the environmental movement would doubtless erode.) Recently, a U.S. district court ruled that the government had to remove the California gnatcatcher, a bird that has been halting development in Southern California, from the list. The court determined that the government had violated the law by not even allowing the public to review the supposedly "best available data."

Subspecies

Another problem is that many federally listed species are not species at all. The Endangered Species Coalition (including groups such as Defenders of Wildlife, the World Wildlife Fund, the Wilderness Society and Audubon) states, "A species consists of those individuals actually or potentially capable of reproducing among themselves but incapable of reproducing with other organisms.

"Populations of a given species from different regions may show characteristic differences in appearance and behavior," the coalition further states. "Such populations are classified as different subspecies or races, but they still belong to the same species." Because the species unit is based upon reproduction rather than slight variations such as blue or brown eyes, its determination is less prone to data errors, bad judgment or abuse by those who would invoke the law for other than its ostensible purpose.

The dusky seaside sparrow, a bird removed from the endangered species list on the grounds that it went extinct, is a perfect example of the subjectivity involved in listing subspecies or distinct populations. The dusky sparrow was considered by some to be a subspecies of the widely distributed and plentiful seaside sparrow. University of Georgia researchers analyzed the last dusky's genetic makeup and concluded that it appeared to be "a routine example of the Atlantic Coast phyla of seaside sparrow" which continues to be plentiful today. Some people proposed back breeding of other sparrows to "reconstitute" a dusky. This may seem strange after just reading that the same people believed it extinct, but those who believe there really ever was something unique called the dusky also believe that it went extinct by hybridizing with other nearly identical subspecies of the same species. What this really shows is that unless we are willing to spend taxes to chaperone wildlife, it is a bad idea to add critters such as the dusky seaside sparrow to the endangered species list.

Costs

In more than two decades, the protection act has effectively "not saved" nonendangered nonspecies. And the costs are beginning to mount. Every time the government adds a plant or animal to the list, it typically produces a plan on how it is going to improve the species status. The plan for the blunt-nosed leopard lizard alone calls for land-use control on or purchases of about 60,000 acres with identifiable costs (many of the planned actions have no cost estimates) exceeding $70 million. More than 800 other species are targets of prepared or pending plans. Then there is a candidate list, or waiting list, of about 4,000 additional plants and animals under consideration. Forty percent of the wildlife on this list are rodents, snails, moths and beetles, including the Tuna Cave cockroach of Puerto Rico—not the "charismatic" species like eagles and bears which are used to pitch the act to the public.

Protection costs are in the billions, but they do not begin to reflect the economic burden imposed on the private sector. Costs for providing unemployment compensation to workers who will lose their jobs if the administration's plan to protect the northern spotted owl is implemented will exceed $700 million annually. It is worth noting that this owl is not really endangered. It was undercounted. It is a subspecies whose close relative is the California spotted owl (which is almost identical genetically) and it is not dependent upon old growth forests for habitat. . . . The government admits that its annual accounting reports on the law "do not reflect a complete picture of the total governmental effort [expenditures]." Although costs have not been accurately measured, even by Washington standards they unquestionably are enormous.

Not surprisingly, many private-property owners are hesitant to provide refuge to federally protected species. When a species is formally declared as endangered, many activities violate the act. This includes clearing dead brush, grazing cattle, putting in erosion barriers, home improvements or anything else that might harm, capture, trap, collect, pursue, harass, wound or kill one federally protected beetle, snail, snake or bird. Each violation carries a possible fine of $25,000 and a year in jail.

Recognizing that conflicts are inevitable and will grow as the federal program increases, Interior Secretary Bruce Babbitt testified in favor of forbidding citizens to use the Freedom of Information Act to find out what federally regulated species are on their property, for fear property owners might rid themselves of the burden. One would hope that a light would go on and bureaucrats would realize that the unintended consequences of their plans are negative for wildlife. Instead, the secretary claims the law provides a means for making plans with property owners so the act will not cause conflict. Sure, if you and your neighbors have a few million dollars to spare and the ability to put up with pounds of paperwork and years of haggling, you might—although chances are slim—get a conservation plan approved. Even then, however, that may not protect you if another conflict with a different animal turns up on the very same property. For the average person, there is no other recourse except to take the government to court, an equally daunting proposition.

Private Interests

It does not need to be this way. Just because this law has been a disaster does not mean that wildlife cannot be effectively managed. There are numerous wildlife success stories, such as those of the wild turkey, bluebird, wood duck, white tailed deer, black bear, beaver, antelope and elk. These animals have rebounded dramatically from severely depleted numbers and did so in large part because of private interests—without the offering of incentives. Just imagine what would be possible if people were rewarded instead of penalized for benefiting endangered species.

The Endangered Species Act is up for reauthorization, and it is going to be a major legislative fight when it makes it to the floor. Fortunately, many members of Congress from both sides of the aisle—such as Reps. Charles Taylor, a Republican from North Carolina; Richard Pombo, a Republican from California; John Doolittle, a Republican from California; Richard Armey, a Republican from Texas; Billy Tauzin, a Democrat from Louisiana; James A. Hayes, a Democrat from Louisiana; and Sens. Larry Craig, a Republican from Idaho, and Richard Shelby, a Democrat from Alabama—are fighting to inject some sanity into the law. Those who earnestly do care for responsible and effective endangered species programs have a serious obligation to address this situation honestly because today's law is based on bad science. It does not save species. It causes conflict, drains resources and may curse any meaningful efforts to help real endangered species—so much so that the entire federal effort is viewed as another nice-sounding government program gone bad.

2 VIEWPOINT

"Two decades of effort to conserve endangered species have produced many successes."

The Endangered Species Act Is Effective

Michael J. Bean

Many endangered species are still in existence due to the Endangered Species Act, Michael J. Bean contends in the following viewpoint. Although Bean admits that these species have yet to be removed from the endangered list, he argues that they have made significant advances in population and have a greater chance to survive than they would have had without the Endangered Species Act. Bean further argues that the act has never received appropriate funding from Congress, a situation that has hampered species recovery programs. Bean is the chairperson of the Environmental Defense Fund's Wildlife Program and the author of *The Evolution of National Wildlife Law*.

As you read, consider the following questions:
1. On what basis does Bean object to the attacks on the scientific data of the federal endangered species programs?
2. According to the author, in what ways did the Endangered Species Act contribute to the recovery of the bald eagle?
3. In Bean's opinion, on what single issue might the critics and the defenders of the Endangered Species Act agree?

Michael J. Bean, "Naysayers Downplay Species Act Successes," *Insight*, May 30, 1994. Reprinted with permission from *Insight*. Copyright 1994 *Insight*. All rights reserved.

In December 1987, Rep. Wilbert J. "Billy" Tauzin, a Democrat from Louisiana, railed away on the House floor against the science underlying the government's endangered species program. Tauzin's ire was directed at regulations requiring fishermen to use gear that was designed specially to halt the drowning of endangered sea turtles in shrimp trawls. The government's estimate that as many as 11,000 sea turtles drowned this way each year was nothing more than a wild extrapolation from limited data, claimed Tauzin. Nowhere near that number of sea turtles could possibly be killed by the shrimp fleet and even if they were, there were better ways of conserving turtles than by requiring the new gear.

Tauzin wanted better science and he got it. The protests of Tauzin and others prompted Congress to ask the National Academy of Sciences to study the issue. Three years later, the academy issued its report: The government's estimate of 11,000 drowned turtles each year probably was wrong, it said; the real number could exceed 50,000. As for a better way of protecting sea turtles, the academy said no—the new gear was essential.

Did the conclusions of the nation's most prestigious scientific body stop Tauzin's attack? Not for a moment. He simply shifted focus from the science to other complaints. His interest in the science, it seems, lasted only until it was clear that the science wouldn't help his real agenda—thwarting the regulations.

Attacks on Science

An even more cynical example of the same phenomenon occurred recently in Alabama. When the Fish and Wildlife Service announced plans to add the Alabama sturgeon to the endangered species list, opponents protested that the proposal was scientifically flawed. They demanded "peer review" by independent scientists. They got it; the reviewers even included a scientist from the Alabama Power Company, a strong opponent of the listing. When word leaked that the review panel's report would unanimously support the service's position, opponents rushed to court to block its release. Technical requirements of the Federal Advisory Committee Act hadn't been met, they claimed. Scientific concerns, once again, had been only a ruse—a tool to gain delay and muster political opposition to a proposal they simply didn't like, regardless of its scientific merits.

These and other examples underscore the point that attacks on the science underlying the Endangered Species Act often are just a smoke screen behind which lurks a fundamental hostility to the law itself. They recall the observation of Father Malagrida, quoted in Stendhal's *The Red and the Black*, that "Man was given the power of speech to enable him to conceal his thoughts."

The delay that Tauzin sought for the regulations protecting sea turtles, the delay that opponents of protecting the Alabama sturgeon secured and the delay that inevitably accompanies similar challenges to the science underlying endangered species decision-making are sure to secure only one result—an increased likelihood of losing the species. And, in fact, it is delay in extending protection to declining species that has given rise to the most serious problems affecting the Endangered Species Act.

Too Little, Too Late

Consider the well-established scientific proposition that, other things being equal, small populations have a higher likelihood of extinction than large ones. Very small populations, in turn, have a very high likelihood of extinction. Disease, inbreeding or even bad luck in the form of freak winter storms can spell the difference between survival and extinction.

The implications for conservation policy are clear: If you wait until a species is reduced to critically low levels before trying to conserve it, you run a high risk of failure. Recovery, if it occurs at all, will be long and expensive. The familiar aphorism that an ounce of prevention is worth a pound of cure is nowhere more applicable.

Yet, waiting until species reach critically low levels before extending them the protection has been the norm rather than the exception. Thirty-eight of the 105 Hawaiian plants on the endangered list had been reduced to 10 or fewer individuals by the time they were added to the list—8 had only a single surviving specimen. According to a recent analysis by David Wilcove and others, half of all the plant species added to the list since 1985 had total known populations of 120 or less. Animals didn't fare much better. The inevitable consequence is that from the outset, the deck is heavily stacked against recovery of most endangered species.

The Road to Recovery

Despite the odds, the remarkable fact is that many endangered species are well on the road to recovery. The nation's symbol, the bald eagle, is a familiar example. Once reduced to fewer than 500 pairs in the lower 48 states, eagle numbers have climbed to nearly 4,000 pairs. In 1994, thousands of people turned out in small towns throughout the Midwest and South in local celebrations of the eagle's recovery.

Naysayers won't credit the Endangered Species Act with contributing to the eagle's recovery. Tauzin insists it was the government's 1972 banning of DDT that should get the credit (though he doesn't dwell long on the acknowledgment that this governmental regulatory action produced dramatic environmental benefits). The Hudson Institute's Dennis Avery won't even that, insisting, "The ban on DDT and the comeback of ~~es~~ is a coincidence." In that view, he is utterly alone and utterly without scientific foundation.

Solid Comebacks

Randall Snodgrass, the National Audubon Society's director of wildlife policy, takes the view that "if we are able to save a single species man has brought to the brink of extinction, then we have been successful."

In fact, the Endangered Species Act's record is even better: Bald eagles, peregrine falcons, and brown pelicans all are making solid comebacks; red wolves have been successfully reintroduced in North Carolina; the California condor and the black-footed ferret, both saved by captive-breeding programs, now are being returned to the wild; and gray whales and sea otters are making a comeback. In the Gulf of Mexico, turtle-excluder devices required on shrimp trawlers are estimated to be saving tens of thousands of endangered sea turtles.

Tom Horton, *Audubon*, March/April 1992.

I wouldn't mind crediting the eagle's recovery entirely to the DDT ban, since my organization [the Environmental Defense Fund] was instrumental in bringing about the ban. But the truth is that while the DDT ban was a necessary first step for the eagle's recovery, it took much more to achieve today's remarkable success. Protection of nesting and roosting habitat, banning the use of lead shot in waterfowl hunting (a source of lead poisoning in eagles), stringent penalties against killing eagles and selling their feathers, restrictions on the popular agricultural pesticide carbofuran and an aggressive program to reintroduce eagles into areas from which they had been extirpated are just a few of the vitally important actions that the Endangered Species Act made possible and without which the eagle's future would be much less secure.

The eagle may be the most familiar of the act's successes, but it is hardly alone. Today, there are more whooping cranes, California sea otters, peregrine falcons, black-footed ferrets, Kirtland's warblers, brown pelicans, red wolves, Aleutian Canada geese and Peter's Mountain mallow wildflowers than at any time in the last quarter-century, thanks to the Endangered Species Act. These are remarkable successes by any reasonable measure, yet because none of these species has yet to recover fully and be removed from the endangered list, naysayers such as Robert Gordon don't credit the efforts as successes. It is like finding fault with a doctor whose patient has been upgraded from critical condition to good because the patient has not yet been discharged.

Lack of Funding

While only two decades of effort to conserve endangered species have produced many successes, much more would have been accomplished if the resources to initiate conservation action earlier had been available. But the truth is that Congress never has provided the funding needed for the act's goals to be accomplished. The $58 million or so that the Fish and Wildlife Service has to implement its endangered species program is roughly what Washington-area residents will spend this year on Domino's pizza.

The consequences of inadequate resources are many: Most species don't make it onto the endangered list until they are in grave peril; the time required for recovery is prolonged; potential conservation options are foreclosed; and remaining recovery strategies often are risky, expensive and controversial. It is no small irony that people such as Gordon who are most critical of the act's record of success have never been heard to argue for the increased resources with which to break this cycle.

Of course, dollars appropriated for the Fish and Wildlife Service are not the act's only costs. Other state and federal agencies and private interests make expenditures and bear costs as well. My advice to readers trying to judge what those costs really are is simple: Be wary. A frequently cited figure is that states spent $131 million in 1992 on behalf of endangered species. But nearly 90 percent of that represented land-acquisition expenditures by a single state—Florida—as part of its "Conservation and Recreation Lands" acquisition program. Because some of the lands acquired harbored endangered species, their cost is included in the above total, even though the lands will be used for recreation and many other purposes.

Playing on the Public's Fears

Outright deception about the economic consequences of endangered species protection is not uncommon among naysayers seeking to stir popular sentiment against that protection. To return to the example of the Alabama sturgeon: When the fish was proposed to be added to the endangered list, Democratic Sen. Richard Shelby and other Alabama politicians couldn't resist playing on the fears of the folks back home who had little knowledge of the Endangered Species Act's actual record of implementation. Shelby and crew ominously warned that if the sturgeon's listing resulted in restrictions on dredging the Tennessee-Tombigbee Waterway, and if those restrictions halted barge traffic on the waterway, then Alabama would suffer a $10 billion hit. All the "ifs" in that statement were correct, but it took the *Birmingham News* to point out that no endangered species anywhere had ever prevented dredging any waterway—not even the pallid sturgeon, a species already on the list. Shelby just couldn't bring himself to tell the voters what a more principled conservative, James J. Kilpatrick, had written only a year earlier: "In practice over the past 19 years, administrators of the Endangered Species Act have quietly exercised the kind of commonsensical judgment that most observers would like to see."

Among voters whose anxieties have been most stirred up by Tauzin, Shelby and Gordon are landowners who believe the administrators of the Endangered Species Act are regularly "taking" private property without the compensation the Constitution guarantees. In fact, however, in the more than 20 years that the Endangered Species Act has been on the books, not one landowner anywhere in the country ever has made a court claim that government action to protect any of the more than 800 endangered species in this country has resulted in property restrictions. When I recently confronted Tauzin with this fact on the PBS program *Technopolitics*, he lamely answered that "it hasn't happened because most people don't have a kangaroo rat in their backyard." Once again, he seeks to instill fear of what might happen rather than understanding of what actually has happened.

Property Rights

On the matter of private property, Tauzin and Shelby apparently don't think the Founding Fathers who drafted the Constitution got it right. They propose to substitute their own standards obligating the government to pay landowners when endangered species and other environmental requirements reduce property values. Tauzin is clear that compensation is not an end in itself, but rather the means of curbing what he regards the excessive zeal of environmental regulators. The "little guy" landowners whose interests Tauzin purports to champion might want to consider that curbing environmental regulation is a double-edged sword. That's a lesson the citizens of Hamilton County, Iowa, won't soon forget.

In April 1994 in Hamilton County, a thousand farmers and others gathered to defend their property rights. They weren't demanding a rollback of environmental laws, however. Instead, they wanted stronger environmental restrictions against the siting of industrial hog lots in their rural communities. Groundwater pollution and the strong odors that only one who has ever been near a huge hog lot can truly appreciate threatened the property values and the quality of life of the assembled farmers. Dennis Figg of the Missouri Department of Conservation tells of being approached by Missouri landowners with similar concerns and being asked to help them find an endangered species with which to defend their property. So much for protecting property rights by crippling the Endangered Species Act.

Incentives for Landowners

The single issue upon which both critics and defenders of the Endangered Species Act might be able to agree is the desirability of creating incentives for better conservation on private land. The act relies heavily on the stick of penalties and prohibitions to deter harmful conduct but fails to offer the carrot of incentives to reward beneficial conduct. In other environmental programs, incentives play important roles. Programs reward farmers for planting erodible land with perennial vegetation and for restoring wetlands; they reward woodlot owners for stewardship of their forest resources; and the Clean Air Act creates incentives for utilities to reduce air pollution more than the law requires. However, we don't have similar programs to reward landowners who could improve habitats and contribute to the recovery of endangered species. We should.

Unfortunately, the zealots leading the attack against the Endangered Species Act show little interest in offering constructive proposals to make it work better; they only want to show how badly it has worked. They have replaced principled conservatism with blind ideology. They seem unaware of the reminder from James L. Buckley, a former senator and federal judge, that it is "a conservative's and a conservationist's instinct to be careful about disturbing systems which seem to have been working reasonably well for an aeon or two." Buckley understood, in a way that Billy Tauzin, Richard Shelby and Robert Gordon seem unable, the significance of 18th-century conservative Edmund Burke's warning that the citizens of any generation are but "temporary possessors and life-rentors" who "should not think it among their rights to cut off the entail, or commit waste on the inheritance" and thereby "leave to those who come after them a ruin instead of a habitation."

3 VIEWPOINT

"Ecological corridors . . . save what is still pristine [and] restore what is still retrievable."

Preserving Ecosystems Will Save Endangered Species

Douglass Lea

In the following viewpoint, Douglass Lea points out the ineffectiveness of attempting to save an endangered species without considering the environment in which the species lives. Arguing that species become endangered as their habitat disappears, Lea suggests that preserving ecosystems is a more efficient method of ensuring the survival of endangered and threatened species. Maintaining a sufficient number of ecosystems can also prevent species from becoming endangered to begin with, Lea asserts. A writer and editor, Lea also teaches in American University's Washington Semester Program.

As you read, consider the following questions:

1. In what ways does fragmentation of habitat threaten plant and animal species, according to Lea?
2. In the author's opinion, how does Ashby's Law of Requisite Variety apply to the preservation of endangered species?
3. What are the benefits of natural corridors, in Lea's opinion?

Douglass Lea, "Linkages and Lifelines," *EPA Journal*, September/October 1992.

"There's a whole lot of death out there," remarks an English conservationist as he watches millions of frogs being annihilated by speeding vehicles. An eight-lane motorway had recently been built across a migration route connecting seasonal poles of the frog's traditional habitat, and the consequences were clear: a massive carnage that left the ancient "frogway" slippery with a shiny soup of crushed amphibians.

On American roads alone, some 100 million wild animals are killed annually. Less dramatic are the steady extinctions of a multitude of obscure flora and fauna, including, at the veiled end of the spectrum, bacteria, fungi, plankton, insects, and mollusks. On a global scale, about 30 million species are thought to exist, and nearly a quarter of them will disappear during the lifetimes of middle-aged human beings.

Loss of Habitat

As the human species spreads into every available niche, wildlife populations become stranded in fragmented islands of habitat, separated from migration routes and normal ranges by roads, fences, dikes, reservoirs, clearcuts, fields of single-crop agriculture, residential developments, and other products of human culture. Often capriciously imposed on the landscape, these overlays of geometry—straight lines, hard edges, acute angles—seldom mirror the natural borders and demands of plant and animal communities. Fragmented habitat isolates these communities, diminishes genetic integrity and viability within species, imperils species that have highly specialized requirements, and encourages exotic and opportunistic species to immigrate and compete for scarce resources.

In the midst of this teeming fertility, pervasive slaughter, and encroaching fragmentation, debate on preserving nature's store of biological diversity, or biodiversity, remains fixed on individual species. Those species with symbolic stature or political utility attract the spotlight. . . . Arguments tend to revolve around the costs and benefits of saving illuminated species like the spotted owl, condor, and snail darter.

True biodiversity occupies a more generous realm. It refers to the full sweep of intricate processes within ecological systems, or ecosystems, and habitats. It provides sufficient redundancy for organisms to adapt to the evolution and shocks in their environment and sufficient variety to resist inbreeding within isolated populations. In *The Diversity of Life*, Harvard professor and prize-winning author E.O. Wilson defines biodiversity as "the variety of organisms considered at all levels, from genetic variants belonging to the same species through arrays of species to arrays of genera, families, and still higher taxonomic levels."

Managing vast arrays of life forms with only the blunt instrument of the Endangered Species Act violates Ashby's Law of Requisite Variety. Derived from cybernetic theory, this law says the repertoire of responses an entity can make to its environment reflects the complexity of that environment. The Law of Requisite Variety implies that a system of strategic controls within the universe of biodiversity-promoting instrumentalities succeeds insofar as it develops a level of complexity similar to that posed by the universe of threats to biodiversity. In these terms, the federal Endangered Species Act, however important in limited applications, constitutes a clumsy response to the complex dilemmas found in the real world of biodiversity.

Ecosystem Management

Three examples show what ecosystem management has accomplished under the Clinton Administration:

A Federal judge in Seattle has approved the Administration's plan for the old-growth forests of the Pacific Northwest. The plan allows for logging to resume at a low enough level that it does not disturb the endangered spotted owl's habitat or the streams that shelter salmon and hundreds of other species.

The State of California has reached an agreement with Federal water and wildlife agencies for managing the San Francisco Bay and its neighboring tributaries and deltas. According to that agreement, water would be allocated to agricultural and urban users in such a way that the spawning grounds of local fish would be protected.

In southern Florida, the Corps of Engineers has begun planning for the rehabilitation of the Everglades, an unfathomably complex ecosystem that in the past 50 years humans have altered almost beyond recognition.

John H. Cushman Jr., *The New York Times*, January 22, 1995.

Fortunately, the variety of responses to those dilemmas evolves more rapidly than the pace and sophistication of legislative process. On a number of fronts—publications, scientific investigations, community projects, school curricula, litigation, state initiatives—the campaign to save the world's biodiversity is beginning to use a wider assortment of techniques and tactics.

Scientific interest has recently focused on mitigating techniques, especially the wildlife corridor, a variation on "greenways," which have appeared in hundreds of local communities and states. Most greenways are designed for human use and enjoyment—typically, an abandoned railroad right-of-way converted to a recreational trail. Wildlife corridors are reserved for plant and animal communities—either to expand ranges and facilitate migrations or encompass shifting habitats under conditions of rapid environmental change, such as global warming. Wildlife corridors serve as "geneways."

Linking Habitats

For example, birds that are unable to survive in a shrunken forest reserve are, nevertheless, able to participate fully in the isolated ecosystem by migrating along forest corridors between reserves. When two or more fragments are linked, the whole is greater than the sum of its parts. In a depleted and simplified exosphere, the mandates of sustainable development, raised to an ethical imperative by the 1992 Earth Summit in Rio de Janeiro, validate the use of coherent networks of ecological corridors to save what is still pristine, restore what is still retrievable, and connect what is still green. To bind remaining wetlands and wildlife reserves, new restorations and nature development areas, and mediating corridors and buffer zones into an extensive system of linear greenways is to create a biological infrastructure for an entire region or country.

Railroad and highway corridors often have biological as well as recreational values. David Burwell, president of the American Rails-to-Trails Conservancy, learned about the potential of wildlife corridors 15 years ago when he received urgent appeals from an official of the South Dakota Wildlife Federation. "He told me the Milwaukee Road Railroad's proposed abandonment of 600 miles of right-of-way would seriously endanger South Dakota pheasants," Burwell recalls. "More than 90 percent of these birds are hatched in the state's railroad and highway corridors. The rest of their habitat has long since been plowed under."

Similar benefits are provided by natural corridors that have suffered relatively little from human occupation or manipulation. In the Southeast, the private purchase of the Pinhook Swamp puts migrating bears out of harm's way and ensures that other plants and wildlife can move freely along a 15-mile corridor between Osceola National Forest in Florida and Okefenookee National Wildlife Refuge in Georgia.

Biodiversity has recently achieved standing on its own merits. In *Marble Mountain Audubon v. Rice*, the Ninth U.S. Circuit Court of Appeals held in September 1990 that a U.S. Forest Service Proposal to log the 3,325-acre Grider Creek watershed had failed to consider its impact on animals using a five-mile-wide corridor between two wilderness areas situated 16 miles apart in northern California's Klamath National Forest. Nathaniel Lawrence, a lawyer for the Natural Resources Defense Council, emphasizes that he argued the case "strictly on the grounds of using corridors to maintain biological diversity and intentionally ignored the menace to threatened and endangered species." This case, in short, transports biodiversity beyond the policy gridlocks forming around the Endangered Species Act.

Meanwhile, bioregionalism, a concept long marginalized at the fringe of the environmental movement, has recently moved to the very center of the biodiversity debate. A California program called Natural Communities Conservation Planning (NCCP) aims to protect critical habitat "before it becomes so fragmented or degraded by development and other use" that its species require listing under an endangered-species program. The program is designed to save critical habitat and, at the same time, allow "reasonable" economic activity and development on affected land, much of which is privately owned. The first NCCP pilot program targets the Coastal Sage Scrub ecosystem, which extends from the Mexican border up the Pacific Coast to Ventura County. Harboring the California gnatcatcher and some 50 other threatened species, this ecosystem demonstrates the advantages of multi-species protection. NCCP's innovations lie in the program's holistic approach to biodiversity and its anticipatory bias—that is, its attempt to stop incipient problems before they become acute and require institutionalized responses.

Experiments in protecting biodiversity find sturdy underpinnings in a growing library of scholarship on the subject and in an expanding number of students learning the principles of conservation biology and landscape ecology. In addition to E.O. Wilson's volumes, the library now includes significant contributions from a wide range of experts, many of whom can be sampled in *Landscape Linkages and Biodiversity*, published in 1991 by Defenders of Wildlife. A comprehensive textbook, *Landscape Ecology* by Richard T.T. Forman and Michel Godron, has been available since 1986. Biodiversity experts have also formed the International Association for Landscape Ecology. Followers of this discipline perform "gap analyses" to generate digital maps that identify both species-rich areas and other ecosystems inadequately protected by existing reserves.

As the frog massacre on the British motorway demonstrates, many species—not just those officially inscribed as endangered species—are now extremely vulnerable. They are at risk from depletion of stratospheric ozone, enormous growth in human populations and economic activity, global warming, droughts, fires, pollution, disease, and other environmental shocks. British authorities, realizing that biological systems require margins of safety, finally tunneled under the motorway to reconstitute the ancient "frogway." Thanks to enlightened management, the frogs are safe again, and their critical role in the maintenance of biological diversity continues.

VIEWPOINT 4

"You can't get three scientists in a room to agree on what an ecosystem is."

Preserving Ecosystems Will Violate Property Rights

Ike C. Sugg

Ike C. Sugg is the Walker Fellow in Environmental Studies at the Competitive Enterprise Institute, a research foundation that promotes the use of private property rights and economic incentives to protect the environment. In the following viewpoint, Sugg argues that ecosystem management—the protection of ecosystems rather than individual species—will enable the federal government to regulate private property and will inhibit development. Because some scientists define an ecosystem as encompassing millions of acres, Sugg asserts, politicians will be able to place large amounts of private property under federal control. Sugg contends that proposals to protect ecosystems are based on faulty science and are primarily motivated by politics rather than by environmental concern.

As you read, consider the following questions:
1. How could ecosystem management enable governmental officials to avoid listing endangered species, in Sugg's opinion?
2. According to Sugg, why would ecosystems be easier to defend politically than endangered species?
3. What does the author mean by the statement, "Owners would become renters on a federal estate"?

From Ike C. Sugg, "Babbitt's Ecobabble," *National Review*, September 20, 1993, ©1993 by National Review, Inc., 150 E. 35th St., New York, NY 10016. Reprinted by permission.

Bruce Babbitt says he was born to be Secretary of the Interior, a destiny that was finally fulfilled in January 1993. Considered by many to be the brightest star in Bill Clinton's Cabinet, Secretary Babbitt is staking his political life on something called "ecosystem management." He views this concept as a way to escape the absolutism of the Endangered Species Act, which has become a regulatory straitjacket for property owners and politicians alike.

Under the ESA, species thought to be in danger of becoming extinct are "listed." Once a species is listed, any human use of land inhabited by that species can be prohibited—"whatever the cost," according to the Supreme Court.

Babbitt's Predicament

From this, it is easy to see why landowners would favor drastic reform, if not outright repeal, of the Endangered Species Act. But to understand why politicians need protection from the ESA, one must appreciate the predicament in which Babbitt finds himself.

Secretary Babbitt is bound by an out-of-court settlement made by the Bush Administration with the Fund for Animals. Under that agreement, the Interior Department must add some four hundred candidate species to the ESA list by September 1996, and expedite consideration for another nine hundred species. Given existing funding levels and the molasses-like speed of Interior's bureaucracy, this would be impossible to do on a species-by-species basis (on average, Interior has been adding about fifty species per year). Even if Babbitt were able to double the annual rate of listing, he would fall far short of the court-ordered mark.

Thus, it is no wonder that Babbitt has proposed to do away with "focusing on single species," opting instead for the "multi-species" ecosystem approach. This will allow "more flexibility" in administering and enforcing the ESA. Rather than going through the laborious process of scientifically documenting the need to list each and every species, Babbitt could simply protect ecosystems that have several hundred species in them.

Essentially, Babbitt wants to avoid having to list species. This is precisely what environmental groups accused Republican Administrations of doing in the past. However, while Republicans may have avoided listings so as to avoid regulating land use, Babbitt is proposing to regulate land use so as to avoid listing species. Cast in Beltway parlance as "proactive," this is what Secretary Babbitt means when he intones that we must "intervene before the crisis." Doing so, he says, will "enhance the position of private property owners."

For a good illustration of how their position is likely to be enhanced, we might take a look at Southern California's Natural

Communities Conservation Plan, which Secretary Babbitt has called a "model" of ecosystem management. Under the plan, developers "voluntarily" idled 200,000 acres of some of the highest-priced real estate in the country, where undeveloped beachfront lots go for as much as $800,000 per acre. Their hope was that such good-faith sacrifice would pre-empt the listing of the California gnatcatcher, a small songbird inhabiting as much as 400,000 acres of the same coastal sage scrub land coveted by developers. Babbitt described the experiment as "breathtaking . . . it's the first time it's ever been done."

Reprinted by permission of Chuck Asay and Creators Syndicate.

If property owners have anything to say, it may well be the last. The gnatcatcher was listed anyway—transforming voluntary efforts into mandatory requirements. Still, Babbitt maintains "it's going to work." According to a puff-piece in *The New York Times Magazine*, Babbitt "brokered a settlement giving protection to the gnatcatcher . . . while still giving the go-ahead to housing developments that had threatened the bird." If this sounds too good to be true, it is. In June 1993, California's Scientific Review Panel released its report, which found that 95 per cent of the coastal sage scrub "ecosystem" would be off limits to development.

It makes little difference to a landowner whether the use of his land is lost through a listing or through "ecosystem management." If ecosystem management will mean regulating private property without waiting for an environmental group to sue Interior for balking on a listing, then it will simply shift the formal cause of regulation from species to ecosystem. Indeed, if the ESA remains unchanged, ecosystem management will prevent neither law suits nor listings. The determining factor will be whether litigious environmental groups believe the existing regulations are adequate; if not, they will use the ESA and sue anyway. If, on the other hand, the regulations are sufficiently stringent to accomplish what the environmentalists would otherwise seek under the ESA, then landowners will be no better off.

The National Biological Survey

The first political hurdle for ecosystem management is passing the $180-million National Biological Survey (NBS) when Congress reconvenes in September 1993. [The funding was subsequently passed, and the agency was later renamed the National Biological Service.] The NBS would be a new Interior Department agency, charged with gathering biological data on public and private lands nationwide.

An important wrinkle is that under the NBS the Freedom of Information Act will be waived. This means that property owners will not have access to the biological information gathered from their own property. They won't be able to challenge the regulation of their land on scientific grounds, as they can now under the ESA. This is akin to saying, "Trust me—your land is part of a very fragile ecosystem, and we don't have to prove it."

This is of more political import than one might think. The fact is that the majority of candidates for ESA listing are invertebrate species, including many insects, arachnids, and mollusks. Babbitt would rather not have to tell property owners they can't use their land because the Delhi Sands fly lives on it. Yet, that is exactly what he may have to tell property owners in San Bernardino County, California, if the fly is listed, which Interior biologists say is a foregone conclusion. [The Delhi Sands fly was subsequently listed as an endangered species.] Politically, ecosystems are much easier to defend than species most Americans wouldn't think twice about stepping on. After all, few people would grasp the ethical theory underlying the ESA's penalty of one year in prison and a $50,000 fine for swatting a Delhi Sands fly. Yet, people may well accept the notion that the fate of humanity depends on that of "ecosystems." After all, as Babbitt says, "Everything's linked." And even if they don't accept that reasoning, the National Biological Survey will ensure that they won't find out why a given area is off limits to human use.

If ecosystems are so important, then it might be worth finding out what one is. According to John Fay, an Interior Department biologist, "You can't get three scientists in a room to agree on what an ecosystem is." One high-level Interior Department official says that "an ecosystem can be an area as small as an acre . . . or can be a very big system such as Yellowstone National Park." In fact, according to the Greater Yellowstone Coalition, the Yellowstone ecosystem is actually much larger than the park itself—encompassing as much as twenty million acres. Earlier, the Coalition had identified that ecosystem as a six-million-acre area. For an ostensibly scientific concept, then, ecosystems are surprisingly subjective. When asked to give a definition at the first hearing on the National Biological Survey in July 1993, Secretary Babbitt responded that an ecosystem "is in the eye of the beholder."

Scary, but true, according to the Forest Service's "National Hierarchical Framework of Ecological Units for Ecosystem Classification," which defines ecosystems as "places where life and environment interact." Clearly, the beholder's eye is looking at an ecosystem whenever it is open.

A Biological Process

Ecosystems are not things, but biological processes in evolution, in which each species plays a role. And society has so altered these processes that we cannot delineate precisely the ecosystem of one species, much less several. We do not know what creatures' evolutionary requirements are, and thus cannot say what habitat they will need over time.

Consequently, a Northwest "old-growth ecosystem" preserve might not protect its resident populations. And unlike individual species recovery plans, it offers no quantifiable way to measure success. Federal accountability, already inadequate, would vanish entirely.

Alston Chase, *The Washington Times*, April 1, 1993.

The legal problem for regulators like Babbitt is that 60 per cent of America's "ecosystems" are privately owned. Perhaps that is why Babbitt talks of "discarding the concept of property and trying to find a different understanding of natural landscape." He wants to do away with "the individualistic view of property," and adopt instead a more "communitarian interpretation," based on the notion that "you can't build fences around property." This is heady stuff.

Historically, government has regulated land use to prevent nuisances. A nuisance is a harm, such as killing the flowers in a neighbor's garden. Simultaneously, government has sought to protect fugitive resources, such as air, water, and wildlife. These transient entities are imbued with a certain "publicness" because they are unowned in the eyes of the law. An ecosystem, however, consists of trees, shrubs, grass, soil, and rocks—the very things of which private property is comprised. If individuals can no longer own the sedentary elements of property, then to what exactly do they have title? In a world run by ecosystem managers, the right to private property would amount to the privilege to pay taxes. Owners would become renters on a federal estate, losing every meaningful incentive to care for the land.

Instead of preventing nuisances, the ESA is forcing landowners to provide public goods. Because wildlife is not privately owned in the U.S., as it can be in parts of Europe and Africa, landowners have no claims against their neighbors' land-use practices that harm resident wildlife. The public does. But the majority of the public's wildlife depends upon privately owned habitat.

This is not, however, an insoluble dilemma. If, as E.O. Wilson of Harvard and Paul Ehrlich of Stanford wrote in a 1991 issue of *Science*, we must "cease developing any more relatively undisturbed land" so as to preserve "bio-diversity," then the public should simply ante up. If the urban majority value wildlife as much as the polls tell us they do, then they should cease compelling the rural minority to provide habitat at their own peril and expense. The Federal Government does not force urban dwellers to grow flowers—yet flowers abound. In fact, they are grown and bought without coercion. Nor do we as a society force defense contractors to build bombs for free. Providing the public good that is habitat for America's wildlife should be similarly viewed—and thus, if anything, should be rewarded, and not punished.

VIEWPOINT 5

"Reintroduction can be worthwhile—and in some cases vital for species survival."

Wildlife Reintroduction Programs Help Endangered Species

Colin Tudge

Zoos and wildlife parks are often the last refuge for endangered species: Some species only exist in captivity, while others maintain larger populations in zoos than in the wild. In the following viewpoint, Colin Tudge argues that endangered species should be reintroduced to their original habitat, whether to increase the numbers of the wild population or to reestablish the presence of a species that has become entirely extinct in the wild. Reintroduction is a difficult process and may take a long time, Tudge concedes, but reintroduction programs already in progress have shown that endangered animals reared in captivity can learn to survive and thrive in the wild. Tudge is a zoologist and a scientific fellow of the Zoological Society of London, England.

As you read, consider the following questions:
1. In Tudge's view, what are the benefits of reintroducing captive animals into wild populations of endangered species?
2. What criteria should be considered when endangered species are being reintroduced, according to Tudge?
3. As described by the author, how might captive-born monkeys help wild-born monkeys at Noah's Park?

Excerpted from *Last Animals at the Zoo* by Colin Tudge. Copyright ©1991 by Colin Tudge. Reprinted by permission of Alexander Hoyt Associates for the publisher, Island Press.

The proper end point of captive breeding is reintroduction. . . .
Probably only a minority of populations bred in captivity will be returned to the wild in the foreseeable future. For some, return may not be possible until many more decades have passed. For others—many, we may hope—mass return may never be necessary, because the wild populations might yet be saved. Wild populations are harder to manage than captive populations, however, and even the 'safe' ones may need genetic support (for future populations are bound to be smaller than those of pre-human times). Neither can we foresee all that fate holds in store for the wilderness of the future, and captive populations should always be capable of return to the wild, if called upon.

Genetic Variation

In reality, too, reintroductions do not have to be absolute, in order to be useful. That is, in the three most famous cases—Père David's deer, Przewalski's horse, and Arabian oryx—the animals were almost certainly extinct in the wild, and survived only because they had been bred in zoos and parks. The ones that were put back were the only non-captive individuals in existence. But reintroduction can be worthwhile—and in some cases vital for species survival—even when the species does still exist in the wild. Thus, golden lion tamarins are being reintroduced into Brazil, and Rothschild's mynah (alias Bali starling) into Indonesia, not to restore an extinct animal but to replenish wild populations that have been dangerously depleted. We may note in this context that captive populations are often more diverse genetically than wild populations of equivalent size, because the captive populations may well have been founded by wild individuals from many different locations, and because nowadays they are bred specifically to maintain genetic diversity. Thus animals introduced to the wild from captivity can be a particularly potent source of genetic variation. To put the matter another way: there may only be a few *species* in captivity that are extinct in the wild; but there is undoubtedly a huge array of *genes* among captive animals that are already extinct in the wild. If those genes are to be returned to the wild populations, then the individuals who carry them must be capable of survival in the wild.

There are many schemes, too, not to reintroduce animals that are totally extinct in the wild, but simply to restore them to parts of their former range where they are now extinct. Thus in Britain, the Royal Society for the Protection of Birds and the Nature Conservancy Council are reintroducing the red kite to England and Scotland. This bird was once common and widespread in Britain. It feeds on carrion and small mammals, and in less hygienic days there was plenty about. London offered especially rich pickings, and the kites (it is said) were as thick in the air as is their equivalent species in present-day Karachi. They declined as street cleaning became fashionable, and gamekeepers began to wage war on them in their inimitably cavalier fashion, until the kites remained in only a few places in upland Wales. They survive in Spain and Scandinavia, however, and are being reintroduced from there. To be sure, this in large part is an exercise simply in 'translocation'; wild animals are simply being moved from one part of their range to a part they no longer occupy. But some birds are being brought in specifically for breeding (including some from a sanctuary in Spain that have been injured, and cannot be released). Thus, captive-bred red kites will swell the ranks of the wild; and thus we see again that reintroduction is not absolute, but can involve many shades of endeavour.

Rescued from Extinction

There are already many species that survive only because self-sustaining captive populations were established before the animals became extinct in the wild. Those species include our own California condor, our black-footed ferret, and our Guam rail, as well as the Arabian oryx, the Przewalski's horse, and the Père David's deer. Their survival in captivity has bought time during which conservation biologists can try to find, improve, and protect suitable habitat for the species to reestablish a wild population.

The Arabian oryx, the California condor, and the Przewalski's horse have already been reintroduced to the wild, and the Guam rail is on its way. Reintroductions are being planned for many other endangered species, as soon as sufficient numbers have been bred in captivity and sufficient wild habitat can be secured. But think of all those other remarkable species that became extinct before captive breeding programs could be established—think of the passenger pigeon, the Carolina parakeet, the Tasmanian wolf, and that symbol of extinction, the dodo. Every such failure to establish a captive breeding program is an inexcusable crime of humans against animals.

Jared Diamond, *Discover*, March 1995.

Reintroduction is never easy, however. Experience gained through various endeavours throughout this century has now enabled the International Union for the Conservation of Nature and Natural Resources (IUCN) to lay down criteria and guidelines. Paramount among them are that the new site should be thoroughly researched. If a particular site last harboured the particular creature 10 years or 100 years, or 1,000 years previously, then, very probably, it will have changed in the intervening years. Does it still provide all the animal needs—nest sites, prey, water, whatever? We know, for example, that the Middle Eastern deserts have been changed radically in recent decades by various 'improvement' schemes, including construction of artesian wells. Is it all still fit for oryx? Primates—orangs, tamarins, woolly monkeys—may be returned to forest; but present forest is likely to be secondary forest (the kind that grows up after primary forest has been felled). Is it suitable?

Even if the territory has not significantly changed since the animal last occupied it, we must ask—why, then, did the animal become extinct in the first place? Do the same dangers that drove it to extinction then, still exist? Two of Britain's reintroduced red kites have already been poisoned by gamekeepers, and although we should still be hopeful (for public opinion is on the side of the birds these days, and those cases led to prosecutions), we have to admit that the initial danger has not entirely gone away. It is indeed vital to ensure (another IUCN stipulation) that the local people are ready to receive the reintroduced animals; and far preferable to find ways in which the local economy (as well as general quality of life) can benefit from their return. Direct benefit to human beings should not be seen as the *point* of conservation; but conservation projects that can bring some benefit are far more likely to succeed. . . .

We would not expect reintroductions to go into full flood until several centuries hence, when the human population has again begun to fall; and we cannot expect that all captive populations can be returned in the foreseeable future, or will need to be returned *en masse*, even in the distant future. But already, despite the caveats and constraints, about 100 reintroductions are under way worldwide, or are already well established. . . .

Learned Behavior

One revelation of the past few decades is that in practice, all but the simplest animals have to *learn* a high proportion of the skills they need for survival; and they do this as humans do, by being placed in the appropriate context at the appropriate age; by trial and error; by observing others; and indeed through direct instruction from their parents, siblings and fellows. Of course—as is also true of human beings—each kind of animal has only a limited repertoire: each kind does tend to do particular things in particular contexts, and in particular ways. But the repertoire is one of potential, rather than of in-built response and activity. . . .

So—if animals are brought up in captivity, they have to learn or relearn the skills of the wild if they are to be returned safely.

A few years ago it would have seemed strange even to contemplate that a captive animal *could* learn its 'natural' skills—and even stranger to suppose that they could learn much of what they need to know from human beings. Indeed it transpires, hardly surprisingly, that some species, some individuals within species, and some age groups, learn more easily than others. In general, though, it is beginning to seem that most captive-raised animals can learn enough at least to get by in a quasi-natural, protected environment; and although very few reintroduced animals have yet gone beyond the first generation in the wild, there is good reason to believe that their offspring, born in the wild, will be able to survive in even wilder conditions. Indeed, we may reasonably hope that as time and the generations pass, reintroduced groups will acquire much or most of the skill and behaviour of their original, wild ancestors—or at least will acquire enough of the ancient knowledge to be able to keep going.

Hence, training captive animals for the wild has become an accepted part of modern conservation practice, taking place before release or after, and preferably both. The general principle—as with all good teaching—is to begin simply and become more complicated; to work step by step from what the pupil knows, to what it does not. . . .

Can captive animals really learn the ways of the wild? The general answer to that general question must be 'yes'. We have already seen it happen hundreds of times. Most instances so far involve domestic animals that have escaped, but there is a growing catalogue of reintroductions for conservational purposes some of which already look convincing. To be sure, some present more problems than others; but problems are there to be overcome.

To date, though, perhaps the most striking vindication of captive breeding comes from Noah's Park in Brazil and the Monkey Sanctuary in Cornwall, Great Britain. Marc van Roosmalen of Noah's Park is seeking to stock his reserve with native, local animals.

These local animals include woolly monkeys. Endangered they may be, but rubber tappers kill the adults for meat, and sell the infants in the local market as pets. One baby woolly is worth a month's wages. The trade is illegal; but intervention seems pointless, once the babies reach the market, because if they were confiscated at that stage there would be nowhere to put them. Noah's Park, however, does provide a suitable sanctuary.

Monkey Tutors

The trouble is that the baby woollies are too young to cope in the wild. By the end of 1990, however, the Monkey Sanctuary had three [captive-born] males of 4 years in age, Nick, Ricky, and Ivan, who were beginning to get to the age when they should lead a group of their own; but the resident dominant male, Charlie, still had many years left in him, and the older Django was established as the number two. For genetic reasons, it seemed that Nick should stay in Cornwall, but the Sanctuary's head keeper, Rachel Hevesi, hatched a plot with Marc van Roosmalen to take Ricky and Ivan out to Noah's Park to act as 'tutors' to the group of babies that had been gathered together in late 1990.

At the time of writing [1991], it is not clear whether the experiment will work. There is every reason to hope that it will, however. For instance, Rachel Hevesi has been able to foster a very young woolly captured from the market on to a still immature female that was already established in the Park. This female took the baby from Rachel, but would return it when called to be fed. Yet this female was not 'tame'; indeed she is one of the Park woollies who never normally comes down from the trees. The general picture is, indeed, that wild-living woollies can establish a relationship with humans that is distant enough to conserve their wildness, but also is close enough to be invoked *ad hoc* for the benefit of the monkeys.

In Cornwall, Ricky and Ivan have already shown that they cope well with infant monkeys, playing with them and allowing them to ride on their backs, in the typical woolly manner. They should be able to impress on the wild-born monkeys the general skills of living, while at the same time, the babies—who were born in the area—should retain good local knowledge of the available food. All the animals should retain enough tolerance of humans to allow themselves to be helped if necessary—but without being dependent on human help. If the experiment succeeds, then its significance will be profound; for this, surely, will be the first occasion on which captive-born animals have been used to tutor and chaperone wild-born animals to help them to survive in their own environment.

Twenty years ago, such an exercise as this would have been considered ridiculous. Now it seems a logical extension of what has already been achieved. That is progress.

6 VIEWPOINT

"Reintroduction . . . is not a panacea for saving animals."

Wildlife Reintroduction Programs Harm Endangered Species

Fiona Sunquist

Most reintroduction programs have failed, Fiona Sunquist contends in the following viewpoint. In some cases, Sunquist asserts, reintroduction attempts have led to injury or death for captive-born animals who were unable to adapt to living in the wild. Furthermore, she argues, those reintroduction programs that have been successful cost far more than other conservation strategies. Noting that the destruction of ecosystems is a major contributor to species extinction, Sunquist suggests that protecting existing habitats would be far more effective than costly reintroduction programs. Sunquist is a frequent contributor to *International Wildlife*.

As you read, consider the following questions:
1. What is the history of the idea of reintroductions, according to Sunquist?
2. In the author's opinion, for what reasons do many reintroduction programs fail?
3. How can adopt-a-park programs save endangered species, in Sunquist's view?

Fiona Sunquist, "Should We Put Them All Back?" *International Wildlife*, September/October 1993. Reprinted with permission.

A helicopter sets down in a clearing in the Sumatran forest, and field workers unload five wooden crates, pry them open and step back. Minutes pass. Finally, the auburn-furred face of an orangutan peers out of one box, staring wide-eyed at the unfamiliar forest.

Tempted by bananas, four other orangs slowly emerge from the crates. Two check out a nearby tree. Another just sits with a gunny sack over its head, slumped in the characteristic boneless orangutan slouch. After watching for an hour or two, the humans depart, leaving the orangs to their new home.

This release took place in 1975. No one knows what happened to the five apes, nor is there much evidence to show that another 300 to 500 orangs turned loose in subsequent years in the forests of Borneo and Sumatra have survived either.

These reintroductions, a touchy issue in Indonesia, came up when orang experts on the World Conservation Union's Species Survival Commission met in January 1993. Their recommendation: Efforts to boost wild orangutan populations had shown "no conservation value" and should stop.

Concerns About Reintroduction

The demise of such a highly visible program raises concerns not just about orangs, but about reintroduction itself. The idea of releasing captive animals to repopulate an area of their original range has become a key strategy for saving endangered species, and people have tried out the idea on everything from Puerto Rican toads and desert pupfish to swift foxes and plains bison. In all, says Ben Beck, associate director for biological programs at the U.S. National Zoo in Washington, D.C., people have attempted 146 reintroductions with 126 different species comprising more than 13 million captive-born creatures, most of them fish. Of the projects on Beck's list, however, only 16 have succeeded.

As scientists review all this work, many are seeing that reintroduction—so boldly touted in the past—is not a panacea for saving animals. Says Devra Kleiman, assistant director for research at the National Zoo and driving force behind releases of the golden lion tamarin in Brazil, "Reintroduction will continue to be part of our conservation repertoire, but it is not a viable option for the majority of species threatened with extinction."

Zoo public relations, the media and just plain wishful thinking may have pumped up unrealistic public expectations. "Reintroductions are seen as examples of man's ultimate ability to solve problems," says Chris Wemmer, director of the National Zoo's Conservation and Research Center. "I think the problem is that we have deceived the public into believing that things are much more simple than they are in the real world."

In all its green, buzzing, crawling, chirping complexity, a real ecosystem has proved not at all easy to restock even with a species it once held. The failures inspire the question: Should more effort go into preventing disaster in the first place?

Early Efforts

The idea of reintroductions arose about the turn of the century with the dawning of the Western conservation ethic. One early attempt involved plains bison, which had been virtually wiped out in the United States by 1890, when their numbers had dropped from 50 million to a few hundred. In 1907, the Bronx Zoo in New York reintroduced 15 captive-bred bison into a reserve in Oklahoma. The animals settled in, established a herd and inspired other bison reintroductions elsewhere. Today the wild bison population numbers more than 100,000.

As human numbers spiraled upward, however, people around the world continued to shoot animals or farm the wild habitats that creatures needed for survival. In some cases, zoos and private animal collectors discovered that they had inadvertently become Noahs tending a modern-day ark as the wild population of a species disappeared, leaving only the captives.

By 1925, for example, the European bison, or wisent, was extinct in the wild. Zoos and private preserves kept the species alive until the 1950s, when breeders released some of the animals into the Bialowieza National Park on the border of Poland and the former U.S.S.R. Now nearly 3,000 wisent roam these deep forests.

By the 1980s, animal-husbandry techniques had so improved that captive populations of some species surged. After 11 generations in captivity, the number of Przewalski's horses topped 1,000, more than zoos could hold. At the Jersey Wildlife Preservation Trust in Britain's Channel Islands, captive populations of the otherwise rare Jamaican hutia, which looks like an oversized guinea pig, began to boom, creating another housing crisis.

Extinct in the Wild

At the same time, threats to additional species led to roundups of the last few animals in disappearing habitats. In recent years, California condors, black-footed ferrets, Puerto Rican parrots and Sumatran rhinos, among perhaps a dozen others, joined the growing list of animals destined for pens as the alternative to extinction.

Once zoos had become proficient enough to breed surplus animals, releasing them back into the wild became more practical. What's more, the idea that the problem of destroying ecosystems could have a happy ending struck a responsive chord with a weary public.

For a few animals—11 percent of attempts, says the National Zoo's Beck—reintroduction has brought a happy ending. He labels a project successful if it creates a self-sustaining population of at least 500 animals. Among the animals on his short list of triumphs are the Canadian wood bison, the alpine ibex in Europe, the bean goose and lesser white-fronted goose in Europe, the fish-eating gharial crocodile in India and the Galápagos iguana and tortoise.

Low Success Rate

Zoos and aquariums in North America house 56 [captive] breeding programs, mostly for endangered or threatened species. Although the efforts are undertaken with the goal of bolstering wild populations, only about 25% of the programs are scheduled to return endangered species to the wild in the next several years. . . .

Many of the animals die when released. A 1989 study published in *Science* magazine found that attempts to put captive-bred species into the wild succeeded only 38% of the time.

Part of the problem is that captivity can change the animal, physiologically and behaviorally. In the wild, there is natural selection: Genes from animals that are best able to survive come to dominate a population. A different sort of selection can occur in captivity: Genetic traits suited to confinement become more dominant.

Maura Dolan, *Los Angeles Times*, May 10, 1991.

One of the best-known triumphs is the Arabian oryx, a desert-dwelling antelope that once roamed much of the Arabian peninsula. The bedouin have long prized both the meat and the trophy head, and the arrival of motor vehicles allowed hunters to chase down antelopes easily. By 1962, only about 30 oryx remained in the wild.

In 1974, the ruler of Oman offered to foot the bill for reintroducing oryx. The first group of captive-bred animals arrived in Oman from San Diego, California, in 1980. Instead of hunting the animals, local Harasis tribesmen have taken on the job of monitoring. By 1991, about 109 oryx were living in the wild.

Vanishing Habitats

For every oryxlike success, there are many creatures that can't go home again. The most common reason is habitat destruction. "Zoos are in real danger of becoming the arks that cannot disembark their cargoes," says Don Lindberg from the San Diego Zoo.

One of the least successful and most controversial reintroduction efforts features the European barn owl. Barn owls are declining throughout Europe, largely because of habitat loss and pesticide use. Britain has only about 4,500 wild pairs, but some 20,000 owls live in private collections and rehabilitation centers. Many captives were crippled in traffic accidents—often birds feeding by the roadside are startled by traffic and fly into cars. Survivors can't fly well enough to hunt, but can still mate. Amateur breeders release an estimated 3,000 captive-bred barn owls each year.

None of the original causes of the bird's decline has changed for the better, and the remaining habitat is already full of wild birds. Most of the released owls either starve within a few months or are hit by vehicles.

Kleiman, Wemmer and other biologists have pointed out that even when bits of habitat remain, reintroduction is a waste of effort unless someone neutralizes the original threat to the species. Though only 2,000 to 3,000 endangered lion-tailed macaques survive in the wild, this silver-and-black monkey breeds so well in captivity that zoos use birth control to prevent overcrowding. Yet conservationists have shelved plans to return captives to India's Western Ghats forests because poachers there still shoot and snare wild macaques. Until enforcement backs up wildlife laws reintroducing these monkeys would benefit outlaws instead of macaques.

Even when habitat is plentiful and poachers are few, captive-bred animals have a tough time adapting to life in the wild, much as a human city slicker would flounder if suddenly dropped into a Sumatran forest.

Animals and birds that migrate long distances, for example, need a wild instructor to teach the route. Scientists tried to solve this problem in whooping cranes' chicks by placing them in nests of sandhill cranes in hopes that the young would follow their surrogate parents on the annual migration. The whoopers did learn to migrate, but they didn't mate, perhaps because they had imprinted on sandhills instead of on their own kind.

Expensive Success

Even with appropriate species, reintroductions take money—big money. This too has raised doubts about the usefulness of reintroduction strategies.

In 1981, Devra Kleiman conceived the now-classic project to return captive-bred golden lion tamarins to the Atlantic forest in Brazil. With colleagues from the United States and Brazil, Kleiman planned the reintroduction as thoroughly as a military exercise, complete with a habitat-restoration plan, education for local people and worldwide public relations. Some tamarins died, some had to be rescued, but others survived and bred. As of 1993, reintroduced tamarins have borne 97 young. Kleiman considered the effort an experiment, and by most standards it has been judged a resounding success. But it was not cheap.

The total money spent to support research and development and field programs in Brazil was approximately $1 million over seven years, Kleiman says. She expects costs to go down as techniques improve, but so far, that's about $22,000 for each surviving tamarin.

The tamarin project may not be unusually expensive; Kleiman and her coworkers are just candid about releasing numbers. The only accounting details available for the oryx project, for example, are that costs run "millions of dollars per year."

Some critics argue that these millions aren't delivering the maximum conservation for the dollar. Ullas Karanth, a biologist with the Centre for Wildlife Studies in India, calculates that $30,000 a year would pay 50 guards to protect enough habitat for 250 to 300 lion-tailed macaques, whereas reintroducing only a dozen would cost $150,000.

Even maintaining captives in the ark costs money. The price tag for keeping viable populations of five species of monkeys alive and well in a zoo is about $500,000 a year. That would pay the total annual costs, including park guards and guides and upkeep, for the whole Serengeti National Park with its vast array of wildlife.

However, the "bring 'em all into the ark" approach still has avid proponents, including Ulysses Seal, chairman of the World Conservation Union's Captive Breeding Specialist Group. He believes that captive breeding and reintroductions are not just important, but essential. "My philosophy is that the ark provides options," says Seal. "If we don't do something now, then in 35 years we will lose somewhere between 1,000 and 2,000 vertebrate species; they won't be around for the options to be tested," he says.

Wemmer is less enthusiastic. "I guess I don't buy into the idea that captive propagation is going to save endangered species as much now as I did in the past," he says.

Preserving Ecosystems

As Kleiman puts it, "It is absurd to think that we can 'save' enough species to recreate an ecosystem like a tropical forest. It is vital that we hold on to the forests and places where these things live in the wild."

A few zoos have already pioneered such projects. The Frankfurt Zoo in Germany, Britain's Jersey Wildlife Preservation Trust and The Wildlife Conservation Society (the research arm of what was once called the Bronx Zoo) have been sponsoring field studies for more than 20 years. Now other zoos are paying for field research. The Metro Toronto Zoo, for example, has sent biologists to study the Liberian mongoose and Nepal's red panda.

Stepping beyond research, the Minnesota Zoo started an adopt-a-park program, pledging $25,000 a year to help support Ujung Kulon National Park in Java, Indonesia. "If every zoo adopted a protected area, we could achieve a significant global conservation program," says Ron Tilson, the Minnesota Zoo's conservation director.

The people of Fort Wayne, Indiana, know it can be done. Earl Wells, director of the Fort Wayne Children's Zoo, convinced the zoo board of directors to designate a year's profit from the gift shop to help protect a species of langur on the tiny island of Mentawai in Indonesia. Then he began a school drive to involve local children. "The response was staggering," says Wells. "Thousands of kids held readathons, sold popcorn and recycled beverage cans—it became the talk of the town." In September 1992, when Indonesian President Suharto promised not to issue any more logging contracts on the island, the children of Fort Wayne sent him 3,500 thank-you letters. Not a bad start for the future of conservation.

VIEWPOINT 7

"We should attempt to protect and preserve every species that needs it, even if we think its survival unlikely."

All Endangered Species Should Be Preserved

Roger L. DiSilvestro

Roger L. DiSilvestro is a conservationist, editor, and the author of numerous books, including *Reclaiming the Last Wild Places*, from which the following viewpoint is excerpted. DiSilvestro maintains that all species are closely interlinked in ways scientists are only beginning to understand. Because seemingly unimportant species may actually be essential, DiSilvestro asserts, it is unwise to purposefully allow species that are unpopular or critically low in numbers to become extinct. DiSilvestro urges conservationists to attempt to save every species possible in order to preserve biodiversity and maintain the natural balance.

As you read, consider the following questions:
1. In what ways are wild plants beneficial to humans, according to DiSilvestro?
2. Why is it impossible to calculate the value of every species, in the author's opinion?
3. In DiSilvestro's view, why is medical triage an inappropriate analogy in the context of species extinction?

Excerpted from *Reclaiming the Last Wild Places: A New Agenda for Biodiversity* by Roger L. DiSilvestro. Copyright ©1993 by John Wiley & Sons, Inc. Reprinted by permission of John Wiley & Sons, Inc.

When conservationists suggest that the planet's biodiversity is decreasing, they are talking about a potential catastrophe for humankind. At worst, rapid loss of species could undercut our own biological supports, stranding us in a denuded world for which we are not adapted and in which we cannot survive. At best, we will almost certainly lose species that could be of critical importance to our well-being and to the improvement of our daily lives. . . .

The Fundamental Things

Even people who have little regard for the spiritual and emotional benefits of biodiversity protection will likely be interested in the fundamental things that wild species bring to our daily lives, such as food, medicine, and even jobs.

"Wild species are in fact both one of the Earth's most important resources and the least utilized," wrote Edward O. Wilson.

> We have come to depend completely on less than 1% of living species for our existence, the remainder waiting untested and fallow. In the course of history . . . we have utilized about 7,000 kinds of plants for food; predominant among these are wheat, rye, maize, and about a dozen other highly domesticated species. Yet there are at least 75,000 edible plants in existence, and many of these are superior to the crop plants in widest use.

You need look no further than the pharmaceutical industry to see another way in which the survival of the natural environment is critical to human society. Since the 1960s, 25 percent of the prescriptions sold in the United States have contained plant extractions as active ingredients. Some 119 medicinal chemicals used worldwide are extracted from plants. These plants number fewer than 90 species, but the planet produces about 250,000 species of plants. The bulk of these have never been tested for their potential value as medicinal substances, and some valuable plant species are disappearing. For example, in the Pacific Northwest, the yew tree was until recently cut down and burned as a trash species. In the 1980s scientists discovered that yew bark contains a chemical effective against some types of cancer. Unfortunately, large quantities of bark are needed to produce useful amounts of the chemical, and too few yew trees of adequate size remain to make the natural source promising. Similarly, in the heavily logged Philippine rainforest some 1,500 plant species are used in traditional medicine. Many of these may offer little more than a placebo effect, but if even a fraction of them are medically useful their extinction would be a major loss to human society.

Wild plants are also valuable because they can help ensure the health or enhance the utility of crop plants. Wild relatives of domestic crop plants sometimes contain genes that give them resistance to threatening diseases. An African species of wild coffee contains a gene that makes it resistant to rust, a disease that in recent years jeopardized the world's coffee crop. Crossbreeding domestic coffees with the wild variety halted the epidemic. A wild grass discovered about 10 years ago in Mexico proved to be a relative of domestic corn. Corn is an expensive crop because it must be grown from seed each year. The Mexican wild grass regrows without reseeding. If its genes for reproductive strategy can be bred into domestic corn, the once-obscure wild grass's potential value—translated into income and jobs—will top $7 billion yearly.

Hidden Worth

It is impossible to calculate the dollar value of every species because complex and indispensable relationships between species compound the value of any given type of plant or animal. A species that seems to have no intrinsic worth might be critical to the survival of another, more clearly valuable species. For example, the Brazil nut is a valuable export crop, bringing in about $1 billion annually to Brazil. But it has never been successfully cultivated—it exists only in the wild. No one knows why it cannot be cultivated, but probably the Brazil nut is in some way dependent on the hidden activities of certain other species, such as insects or rodents. Thus the billion-dollar Brazil-nut industry might depend ultimately on the fate of some unknown moth.

Hope for Survival

Scientists say that a population bottleneck is not a death sentence. In the past, small groups of inbred animals have recovered. The elephant seal was down to about 20 individuals off Mexico and California at the beginning of the twentieth century, for instance. But after protective legislation, the numbers rebounded until today the huge beasts frolic by the tens of thousands.

"Even if there are only a few animals left, there is no such thing as a lost cause," argues population biologist Michael Soule. "There is only the loss of hope and the shortage of funds. With enough hope, financial support and proper management, we can save almost anything."

Sharon Begley, *National Wildlife*, February/March 1991.

Even a species of obvious value to humans may have greater value to its ecosystem. The African elephant, for example, is critically related to a wide number of other species. During droughts, elephants dig waterholes in river beds, providing moisture for uncounted other species. The seeds of some trees, such as the acacia, germinate best when they have passed through an elephant's digestive tract. Elephants knock down mature trees, helping to keep grasslands open—a critical factor for many grazing species, such as antelopes. The elephant's value is thus more than the price of its tusks or the money it brings from tourism. It is the value of nearly all the large animals that share the elephant's habitat.

The mutual dependence of species makes the protection of biodiversity critical. As we remove species from any ecosystem, we begin to upset the balance of all species. Remove enough species, or a keystone species such as the elephant, and the whole system veers toward collapse.

Underlying the survival of all species is habitat itself. Without proper habitat—whether it be virgin forest, open grassland, the bottom of the sea, or the human intestine—no species can survive. Habitat protection, which includes the protection of everything from wilderness areas to city parks, is the foundation of all efforts to protect species, to preserve biodiversity. This is why it is so vital to save such vanishing habitats as the temperate rainforests of the Pacific Northwest and the tall-grass prairies of the Midwest. They represent the last of a certain type of ecosystem. When they are gone, we risk losing all that these ecosystems offer us. This is equally true of wetlands and coastal zones. The loss of these habitats results in a net loss of species, a diminishing of biodiversity and all that it promises us in foods, medicines, and other products. . . .

The Importance of Low-Profile Species

The use of popular, attractive endangered species as drawing cards for public support tends to yield conservation programs that focus heavily on popular, attractive endangered species. As a result, the federal government spends millions of dollars yearly on such high-profile species as the gray wolf, grizzly bear, and California condor. Meanwhile, less appealing but equally important species—usually bloodless, cold-to-the-touch, but biologically significant species that include dozens of insects, shellfish, and reptiles—are ignored.

To a biologist, of course, the protection of the endangered Cumberland monkeyface pearly mussel—a real species on the federal list—is equal in importance to the protection of the California condor. A biologist knows that the streams of North America were once crowded with clams and mussels. Early settlers in Ohio reported freshwater clams as big as dinner plates that paved the bottoms of clear-running streams. But the streams no longer run clear, and many of the shellfish species have disappeared or become endangered. Their decline is not only a sign of ecological disintegration, but also a loss of a vital riparian component, for shellfish helped filter debris from the streams. Their loss compounds the damage European settlement has done to our waterways.

Nevertheless, public support is far easier to rally for big and impressive creatures. Logically, therefore, the conservation community focuses on the most appealing species, often ignoring the bigger question of ecosystem protection. Since the private groups generally play the leading role in keeping government land-management agencies abreast of the times, it is critical that the groups begin to emphasize the need for ecosystem rather than single-species management. At least one organization, Defenders of Wildlife, has undertaken to put biodiversity protection and ecosystem management on the national agenda, and presumably others will soon join the effort.

Some conservationists, adamant about the need for ecosystems management, want to scrap the endangered-species program or at least give up on management-intensive, and expensive, species such as the California condor. This faction of the conservation community urges the Fish and Wildlife Service to apply medical triage to endangered species protection. Medical triage is used during times of war when medical resources are scarce. Only patients likely to respond to treatment are given aid. No treatment is given to patients likely to survive without treatment or likely to die without a massive effort. With triage in mind, some conservationists suggest that seemingly hopeless species—and no one has ever defined this term—be left to extinction.

Of course this is ludicrous. We can never, with the certainty of medical science, predict which species will fail and which will not. Moreover, speaking objectively, we have an apples and oranges analogy here. Triage advocates are comparing the loss of an individual to the loss of a species. But the loss of an individual member of a species is not biologically significant. The loss of an entire species is.

Playing God

We cannot play God by selectively wiping out species through inaction (though when it comes to playing God, we always seem comfortable with destroying life and taxed by nurturing it). Each lost species represents a decline in local and global biodiversity, and it is this decline that all the works and endeavors of conservation seek to avoid. We should attempt to protect and preserve every species that needs it, even if we think its survival unlikely. We lack the knowledge to determine which species will inevitably fail and which we can save. Better to err on the side of life than on the side of oblivion. Moreover, we should not pass on to future generations a diminished biodiversity. They may have skills and abilities yet undreamed, and it behooves us to ensure that they do not have less to work with than we had.

Rather than draw analogies with medical triage, let us seek metaphors in the legal system. The funds we expend on endangered-species management are the fine we are required to pay for past crimes and misdemeanors against nature. The result will be that future generations can, perhaps, gaze into the skies of California—and perhaps of Oregon and Washington State—and see soaring on whistling wings the California condors that our efforts in behalf of the endangered have bequeathed to them. Isn't this a nobler goal than letting the largest bird on the continent vanish, leaving future generations to wonder why we deprived them of this marvel?

The conservation community should not waste time providing the world with justifications for discarding endangered wildlife. It should focus on finding more funds for biodiversity protection and more means for bioeducating the world, creating a place where wildlife and wildlands are better understood and more highly valued.

VIEWPOINT 8

> *"[Saving] the most-endangered species . . . diverts attention and money from the much more crucial goal of preserving overall biological diversity."*

Some Endangered Species Cannot Be Saved

Suzanne Winckler

Going to extreme measures in attempts to save endangered species with little hope of recovery is a mistake, Suzanne Winckler argues in the following viewpoint. Such desperate remedies—especially as implemented under the Endangered Species Act—take funds and aid away from those endangered species with more potential for survival, Winckler maintains. A better approach, she contends, would be to preserve those ecosystems that contain the widest amount of biological diversity. Winckler is the author of *Great Lakes* and *Plains States*.

As you read, consider the following questions:

1. What objections does Winckler make to the refusal of conservationists to "play God"?
2. Why do people prefer to help critically endangered species, in the author's opinion?
3. According to Winckler, how does the Endangered Species Act block the conservation of ecosystems?

Suzanne Winckler, "Stopgap Measures," *The Atlantic Monthly*, January 1992. Reprinted with permission.

To say that the Endangered Species Act is not working is to sound ungrateful for what it has accomplished. Inasmuch as tens, if not hundreds, of organisms that would undoubtedly be extinct by now—including the Attwater's prairie chicken, the Florida panther, the black-footed ferret, the Kirtland's warbler, and the Puerto Rican parrot—are instead hanging on by a thread, the act has been a success.

The continued existence, however precarious, of these species is deeply satisfying to many people. The creatures are beautiful (the whooping crane); they stand for ideals that are important to us culturally (the bald eagle); they exhibit incredible behavior (the Attwater's prairie chicken); they are stunning emblems of the closest we can come to pristine wilderness (the grizzly bear). Our knowledge of their presence in the wild helps assuage our guilt about what we've done to them in particular and to the natural world in general, which might imply that the Endangered Species Act is ultimately designed to treat our own brand of sickness and not theirs. Regardless of what these animals do to make us feel better, they are the walking wounded of the world, and it costs millions of dollars to keep them out there.

Faint Praise

That it saves organisms from extinction is faint praise for a law with the far loftier aspiration of "better safeguarding, for the benefit of all citizens, the Nation's heritage in fish, wildlife, and plants." It is the stated purpose of the law not just to keep species from going extinct but to return them to viability. In this regard it is failing.

The Endangered Species Act takes under its wing an array of taxa—from full species (whooping crane) to subspecies (Attwater's prairie chicken) to discrete populations of species (the Mojave population of the desert tortoise). "Taxon" (plural "taxa") refers to any of the groupings into which taxonomists classify organisms. As of November 1991 the federal government listed 1,196 taxa around the world—more than half of them occurring in the United States and its territories—as either endangered or threatened, one of which dubious distinctions is necessary for care under the act. Another 3,500 or so are waiting for review. Among this second group—known as candidate species or Category 1 and Category 2 species—are many plants and animals acknowledged by scientists to be in far worse danger than species that have already qualified. They languish in bureaucratic limbo because of a perennial problem of the act: it never has enough funding. A number of candidate species have gone extinct before they could be considered for listing.

The process for listing threatened and endangered species is complicated, but if there is an overriding criterion for listing, it is that the species is demonstrably imperiled. John Fay, a botanist with the Division of Endangered Species of the U.S. Fish and Wildlife Service, explains the implicit ranking process: "We try to set our priorities so that those species that face the greatest threat are the ones we address first. The alternative—intervening with things that are in better shape—would mean losing a substantial number of species in immediate danger. And within the terms of the Endangered Species Act that's unacceptable."

Triage

This is what critics point to as a major failing of the act. It intervenes in a way that no intelligent nurse, paramedic, or doctor would under analogous circumstances. It has thrown out the window any concept of triage. It does not sort and care for species in such a way as to maximize the number of surviving species. On the contrary, it attempts to save the hardest cases, the equivalent of the terminally ill and the brain-dead. It pays less attention to species that would be easier and cheaper to save—species that require treatment akin to minor surgery, a splint, or a Band-Aid. The act has no concept of preventive medicine, of keeping healthy species from peril. Consequently, many animals and plants that were common twenty years ago—were even considered in some realms to be pests—are now entitled to care under the Endangered Species Act.

Cost-Conscious Noahs

Few species are unsavable today; concerted human effort can save most of them. But we are unlikely to have the means to save them all. In this deficit-ridden age Fish and Wildlife Service budgets will not climb to the altitude necessary to save the few hundred species on the list, let alone the thousands upon thousands of unlisted species that biologists regard as endangered. Like cost-conscious Noahs, Americans will pick which creatures to bring with them and which to leave behind. The choice is inescapable.

Charles C. Mann and Mark L. Plummer, *The Atlantic Monthly*, January 1992.

One reason the act eschews triage is that its enforcers, not to mention many endangered-species watchdogs, do not want—and do not want to allow anyone else—to "play God." There is no doubt that such a role confers awesome responsibilities. The triage of species and the triage of individuals of one species (as it is practiced in emergency rooms and on battlefields) differ by several orders of magnitude. Few people would want the ethical burden of deciding the fate of a whole species—of saying, for instance, that the blunt-nosed leopard lizard can go but the humpback whale stays. John Fay says, "All we can do is try to preserve our options. My absolutely favorite quote is from Aldo Leopold: 'To save every cog in the wheel is the first precaution of intelligent tinkering.' That's what we're trying to do. We don't know which ones are important. We don't know which ones are going to disappear."

Moral neutrality is noble, but it creates problems in the categories of money (there has never been enough allocated to save every cog; there is no promise of more in the future) and biology (the cogs we are saving are so crippled, so compromised, that they can barely perform their assigned functions in their respective niches). To refuse to play God is to play the devil by default.

Crisis Management

It's also true that saving species that don't need much fixing is boring. We thrive on crisis management, and we love things that are rare much more than things that are common. The people in the public and private sectors who work on policy that actually attempts to protect species before they become endangered—for instance, the Fish and Wildlife Service's Office of Migratory Bird Management, the Nature Conservancy, and the International Council for Bird Preservation—receive precious little attention from the media for their work, because what they do is tedious and unglamorous: it does not play well in measured media doses the way a hand puppet feeding a nestling condor does.

In 1988 Congress began to require that the Fish and Wildlife Service submit annual reports on federal expenditures for the U.S. roster of endangered and threatened species. (The federal government spends little money on foreign species.) This is not an easy task—at least thirteen federal agencies make regular outlays of money for endangered species. Nor are the reports ever likely to be more than a best guess of expenditures, since endangered-species activities often merge with other operations. For example, prescribed burning—setting fires to remove the brushy vegetation once held at bay by natural fires—is a management tool at Francis Marion National Forest, in coastal South Carolina, which aids not only endangered red-cockaded woodpeckers but also game species like turkey and quail.

While there is much to be said for making the protectors of endangered species accountable for how they spend our money, the expense reports are sadly divisive documents that reinforce the individual-species thrust of the Endangered Species Act. They provide ammunition not only for those who are alarmed by how much we're spending on, say, the Higgins' eye pearly mussel ($437,700 in 1989; $367,000 in 1990) but also for those who are upset at what we're *not* spending on, say, the desert tortoise, a species that because of a strange upper-respiratory disease is dropping dead at an alarming pace in the Mojave Desert. The tortoise got almost $500,000 in 1989. Its advocates saw that boosted to more than $4 million in 1990.

The Big Picture

By focusing on individual species, the expenditure reports perpetuate a chronic lack of attention to the big picture. Environmentalists are defending expenditures for the plants and animals to which they have, for whatever reason, chosen allegiance when instead they should be addressing a whole different set of questions and concerns. Why does the list of endangered and threatened species keep getting longer? Why have only a few species ever been taken off the list? Where on earth will the money come from to care for every new addition to the lengthening list?

The biggest question the expenditure reports should provoke is this: Why is half of all the money earmarked for endangered and threatened species being spent on only twelve of them? In 1990, of $102 million apportioned among 591 taxa, a total of $55 million went to twelve. They are, in descending order of expenditures, the northern spotted owl, the least Bell's vireo, the grizzly bear, the red-cockaded woodpecker, the Florida panther, the desert tortoise, the bald eagle, the ocelot, the jaguarundi, the peregrine falcon, the California least tern, and the Chinook salmon. The apportionment does not become much more equal after this first dozen. The next dozen species—which include the gray wolf, the southern sea otter, and the Puerto Rican parrot—received the next $19 million. In other words, the remaining quarter of funding—$28 million—was shared among about 570 other organisms.

The fortunate two dozen or so creatures that command three quarters of the money are among the most beautiful on earth. They have captured the hearts of a wide assortment of people. These people would better serve the objects of their affections if they cared for whole ecosystems—the degradation of which is largely responsible for our degraded wildlife—with the same fervor.

The Cost of Intensive Care

Intensive care for animals and plants is costly. It is not cheap to hire airplanes and helicopters for the surveillance of populations (as was done before the last California condors were taken from the wild), set up captive-breeding facilities (for the bald eagle, the peregrine falcon, the whooping crane, and the black-footed ferret), translocate animals, either in order to get them away from the threat of people or in order to invigorate isolated gene pools (the grizzly bear, the red-cockaded woodpecker), release individuals back into the wild (successfully accomplished with the peregrine falcon), attempt to establish breeding populations (tried and failed with the whooping crane), or keep parasitic cowbirds out of the nests of endangered birds (being done year in and year out for the Kirtland's warbler, the golden-cheeked warbler, and the black-capped vireo). When we focus treatment on the most afflicted, we can expect to pay.

It is true that the Endangered Species Act inherited some desperate cases when it went into effect. The whooping crane, the California condor, and the Kirtland's warbler are examples of species that were rare to begin with and suffered from the presence of people within their ranges. Thirty years before the Endangered Species Act, biologists had begun the arduous and expensive endeavor of bringing the whooping crane back, from a single flock of fourteen individuals in the wild. After fifty years of intensive management, the wild flock of whooping cranes numbers about 140 birds. This population—like all small populations—is intensely vulnerable: last winter [1989-1990] nine birds, representing six percent of the flock, were lost.

It is also true that many endangered and threatened species need only moderate sums of money to survive. For instance, certain narrowly endemic plants—plants that have evolved in rare microhabitats, such as the bunched cory cactus and the McKittrick pennyroyal—require little more than the purchase of the land on which they grow, and small outlays for monitoring and law enforcement (cacti, for example, are particularly susceptible to rapacious collectors).

But for every one of these bargain species there is a very expensive one waiting for a constituency to care enough, waiting to get a little more endangered, or waiting for its recovery plan to be approved. . . .

Saving the Majority

In order to save the most-endangered species, the act diverts attention and money from the much more crucial goal of preserving overall biological diversity—that is, preserving the maximum number of healthy species in ecosystems that require a minimum of maintenance. The way to save species is to save the places where they live. By extension, the way to save the greatest number of species is to save the places that house the richest biological inventory. . . .

The Endangered Species Act has institutionalized the bizarre notion that the primary legal justification for the preservation of an ecosystem is a species teetering on the brink of extinction. That it is one of the most magnificent landscapes in North America is somehow no longer reason enough to preserve the last remnants of the Pacific old-growth forest. Instead, the only legal mechanism available is to require the preservation of some minimum configuration of that forest in hopes of keeping the northern spotted owl—one species among thousands that dwell there—from going extinct. At the same time, the boreal forests of Minnesota, Wisconsin, and upper Michigan, the marshes and wetlands rimming our coasts, the prairie potholes of the Midwest, the riparian woodlands along streams in the West, and other ecosystems will continue to shrink until they yield evidence of the endangered species that will warrant their preservation. . . .

Perhaps the evolved policies of the Endangered Species Act are the best we can hope for in an imperfect world. But when I think of biologists frantically building nest holes for red-cockaded woodpeckers, or keeping vigil under the last few Puerto Rican parrot nests in the wild, or watching black-footed ferrets die of canine distemper, or abandoning their efforts to establish another flock of whooping cranes at Grays Lake, Idaho, or pitching cowbirds out of the nests of Kirtland's warblers year in and year out, I no longer call to mind the words of Aldo Leopold or Henry David Thoreau or John Muir or any of the Native American chiefs who spoke so eloquently long ago about the sacredness of the earth and mankind's debt to the beasts. Instead, I think of Hampton Carson, a geneticist and an authority on endangered Hawaiian flora and fauna, who once wrote, "Nature is a better stockkeeper than we are."

Periodical Bibliography

The following articles have been selected to supplement the diverse views presented in this chapter.

William O. Briggs Jr.	"The Endangered Species Act Must Be Changed," *21st Century Science and Technology*, Fall 1992. Available from PO Box 16285, Washington, DC 20041.
Douglas H. Chadwick	"Dead or Alive: The Endangered Species Act," *National Geographic*, March 1994.
Andrew Neal Cohen	"Weeding the Garden," *The Atlantic Monthly*, November 1992.
John H. Cushman Jr.	"Timber! A New Idea Is Crashing," *The New York Times*, January 22, 1995.
Jared Diamond	"Playing God at the Zoo," *Discover*, March 1995.
Gerald Leape	"We Need Laws to Preserve Biodiversity," *The World & I*, September 1993. Available from 2800 New York Ave. NE, Washington, DC 20002.
Stephen M. Meyer	"The Final Act," *The New Republic*, August 15, 1994.
Micah Morrison	"Babbitt's Gambit: Interior Designs," *Insight*, August 3, 1993. Available from 3600 New York Ave. NE, Washington, DC 20002.
Charles Oliver	"All Creatures Great and Small," *Reason*, April 1992.
Peter Radetsky	"Back to Nature," *Discover*, July 1993.
Walter V. Reid	"Toward a National Biodiversity Policy," *Issues in Science and Technology*, Spring 1992.
Rodger Schlickeisen	"A Better Approach to Preserving Species," *The Christian Science Monitor*, October 22, 1992. Available from One Norway St., Boston, MA 02115.
Nancy Shute	"Thinking Big," *The Amicus Journal*, Fall 1992.
Kenneth Smith	"Save-the-Roach Campaign Bugs at Least One Taxpayer," *Insight*, June 1, 1992.

CHAPTER 3

Should Endangered Species Take Priority over Jobs, Development, and Property Rights?

Endangered Species

Chapter Preface

In 1972 Eric Forsman, a graduate student at Oregon State University, first warned that the northern spotted owl population was "declining as a result of habitat loss" due to logging. The owl was among the first candidates for listing under the 1973 Endangered Species Act, which would provide federal protection against activities that could harm the species' chances of survival. However, in 1981 the Fish and Wildlife Service (FWS) decided not to list the northern spotted owl as an endangered species. Due to this decision, environmental groups sued the FWS, the Forest Service, and the Bureau of Land Management in the late 1980s. These lawsuits resulted in the suspension of most logging on federal land in the Pacific Northwest from 1989 to 1994.

Timber industry executives, loggers, and sawmill workers fiercely opposed the logging ban, which they blamed for the drastic decline of forest industry jobs that occurred in the late 1980s and early 1990s. Protesters carried signs that read "Save a logger—kill an owl" and "Loggers are an endangered species." Nevertheless, on June 22, 1990, the FWS officially listed the northern spotted owl as a threatened species under the Endangered Species Act. In September 1991, then–Secretary of the Interior Manuel Lujan convened the Endangered Species Committee (popularly referred to as the "God Squad") to determine if the listing of the northern spotted owl could be revoked due to "severe economic hardship." Although the committee voted to allow logging on several federal timber tracts previously affected by the ban, the God Squad ultimately decided not to overrule the FWS's decision to list the owl.

On April 13, 1994, the Clinton administration laid out its proposed solution to the spotted owl controversy, the Pacific Northwest Forest Plan, which would set aside more than 16 million acres of old-growth forest for the spotted owl while reopening about 5.5 million acres for logging. The administration estimates that approximately 9,500 lumber jobs will be lost under this plan, while timber industry executives place the estimate at 85,000 jobs. Timber industry organizations and environmental groups both responded to the new plan by preparing lawsuits to overturn it: The timber industry argues that the plan will result in unacceptable levels of unemployment, while the environmentalists claim that the amount of protected acreage is too small to sufficiently protect the northern spotted owl.

The ongoing debate divides sharply between proponents of saving the northern spotted owl and proponents of preserving loggers' jobs. The controversy over the fate of the threatened spotted owl and of the loggers' endangered way of life is one of the issues discussed in the following chapter.

VIEWPOINT 1

"The [Endangered Species Act] is accomplishing precisely what it was intended to: the uncompensated taking of private property."

Protection of Endangered Species Harms Private Property Owners

Robert J. Smith

Under the Endangered Species Act, the federal government has the power to protect an endangered species' habitat even if that habitat is on privately owned land. In the following viewpoint, Robert J. Smith argues that, due to this policy, farmers and other private property owners have lost thousands of dollars in income and real estate value. It is unconstitutional to restrict private owners from farming or developing their land in order to protect endangered species without compensating the owners for their financial losses, Smith asserts. These harsh policies, Smith maintains, also lead landowners to "shoot, shovel, and shut up" about any endangered species they might find on their property. Smith is a senior environmental scholar at the Competitive Enterprise Institute in Washington, D.C.

As you read, consider the following questions:

1. How did government actions lead to the destruction of kangaroo rat habitat, according to Smith?
2. What two amendments to the Endangered Species Act does Smith suggest?

Robert J. Smith, "Fire, Rats, and the Endangered Species Act," *Regulation*, vol. 16, no. 4 (1994), pp. 14-16. Reprinted by permission of the Cato Institute.

Cindy and Andy Domenigoni are fifth generation farmers in Riverside County, California, working a 3,200-acre property first farmed by Andy's great-great-grandfather who settled the valley (that is now named after him) in 1879. Their farm has also been home to the Stephens' kangaroo rat (k-rat), a species the government has listed as endangered since 1988. In compliance with the Endangered Species Act (ESA), the Fish and Wildlife Service (FWS) has prohibited the Domenigonis from farming 800 tillable acres that are considered prime rat habitat.

In 1990, as the Domenigonis were preparing to begin plowing their fields, FWS law enforcement agents and biologists ordered them to stop and warned them that disking their fields would constitute a "taking" of the endangered Stephens' kangaroo rat and they would be arrested. Furthermore, they were cautioned that if they subsequently disked their fields, they would face impoundment of their farm equipment and a year in jail or a $50,000 fine—or both—for each and every act of "taking" an individual rat. And as the FWS considers a taking to mean harassment, harm, digging up a burrow, or plowing under the grass and plants whose seeds the rat eats—almost any action a bureaucrat can conceive of as affecting the rat in any way—the Domenigonis could have been facing life sentences for plowing their 800 acres. (That is what the environmentalists refer to as "sustainable development.")

Financial Hardship

By 1994 the Domenigonis had lost $75,000 in forgone crops each season—a total loss of $300,000 in gross income—because of the FWS prohibition. They had also incurred another $100,000 in biological consulting fees, legal fees, and associated costs in fighting this regulatory taking of their property and of their livelihood. In addition, they have been prevented from raising crops on other farmland that they leased from local landowners.

Ironically, on November 1, 1993, shortly after the devastating Southern California fires destroyed thousands of acres of k-rat habitat—as well as human habitat and homes—FWS biologist John Bradley authorized the Domenigonis to plow their fields, having determined that the endangered rats no longer lived in the area. However, it was not because the fires had burned the rat habitat (i.e., the fallow fields) to a crisp, along with the rats. Biologist Bradley said the k-rats had already left the area before the fire because the years of leaving the fields fallow had made the brush and weeds grow too thick for them!

Thus, the ESA regulations directly caused an uncompensated loss to the Domenigonis of close to half a million dollars. Their land had undergone a de facto nationalization by the federal government; they could derive no economic return from it. Yet they were still required to pay property taxes on land deprived of all economic value by government fiat. Furthermore, the government's actions apparently led to the destruction of numbers of the endangered Stephens' kangaroo rat and large areas of its critical habitat.

Compensating Property Owners Helps Species

Fearful of seeing their land thrown into limbo by the U.S. Fish and Wildlife Service, many property owners regard protected species with hostility. According to Terry Anderson, a PERC [Political Economy Research Center] senior associate, "Some farmers shoot, shovel and shut up," when they see an endangered species on their land.

Defenders of Wildlife in Washington, D.C., has a far better idea. It launched a $100,000 fund to compensate ranchers for any livestock killed by at-risk gray wolves in Montana. Since 1987, the fund has paid some $12,000 to a dozen Montanans who suffered wolf-related animal losses. In 1992, the group created the Wolf Reward Program which pays $5,000 each to landowners who successfully have wolves breed and raise their pups to adulthood on their private land.

Deroy Murdock, *The San Diego Union-Tribune*, February 11, 1994.

Some FWS spin doctors are now arguing that the fires helped the rats because they don't like thick brush and now preferred plants will take over. They also point out that there are substantial numbers of unburned seeds in the ashes and debris. However, untold numbers of rats perished in the fire, whether caught out at night or baked in their burrows. While seeds may be present now, the winter rains will soon wash them away and erosion, sedimentation, floods, and mudslides will eradicate still more rats and their burrows. Although they may not prefer thick cover, the absence of any cover will make them more susceptible to predation by birds of prey. There is also the delicate question of what the k-rats will eat until all this new preferred vegetation grows in over the coming years and finally produces seed crops. Will we see . . . preternaturally intelligent rats gathering on hilltops waiting for FWS airdrops of k-rat chow?

Federal "Protection"

The question has been raised that if the Domenigoni spread, which had been farmed for well over a century, constituted prime Stephens' kangaroo rat habitat, then how bad could farming have been in the first place? Indeed, as soon as the FWS biologists banned farming in the rat habitat, first the rats disappeared because their habitat was no longer any good, and then it all burned to the ground anyway. The k-rat breeds new generations rapidly; perhaps it had little problem adapting to dryland farming, disking, pest control, and all the activities that the Feds prohibited. Considering that the k-rat had existed for countless centuries before Europeans arrived, and had at least coexisted for over a century with ever more modern agricultural technologies, it is highly ironic that the rat vanished after a mere five years of caring federal "protection."

This travesty of justice; this violation of the constitutional rights of American citizens—specifically of the Fifth Amendment's takings clause, which declares "nor shall private property be taken for public use without just compensation"; this destruction of a federally endangered species and its habitat—through the enforcement of the ESA—is repeated over and over with examples of species after species from all across the United States. Yet the environmentalists and the Department of the Interior continue to proclaim the ESA a successful, pragmatic, and wonderful law. That requires a considerable stretch of the imagination. Or as a growing host of critics of the ESA with first-hand knowledge of its workings would attest, it demonstrates that the ESA is accomplishing precisely what it was intended to: the uncompensated taking of private property and control of economic growth, development, and "urban sprawl" through de facto national land-use control—all without the necessity of paying any form of compensation.

It is possible to have both a free and prosperous society and a sound and healthy environment. Much of American history attests to this. But first it requires secure property rights and positive incentives for landowners. The ESA provides neither and turns the experience and hope of private stewardship from a win-win situation into a lose-lose situation.

Save a Rat, Lose a Home

The brutal realities of the ESA were exhibited to the entire nation on ABC's *20/20* television news program of Friday, November 19, 1993 (hosted by Hugh Downs and Barbara Walters and reported by John Stossel), where Ms. Anna Klimko, who obeyed the federal government's orders not to create a firebreak by plowing the brush in front of her house because doing so would damage the k-rat's burrows and therefore harm the k-rat, was kneeling in the ashes of her totally destroyed home and dreams, digging for the possible remnants of family keepsakes. Ms. Klimko looked up with tears streaming down her face and asked, "In three minutes, my house was fully consumed in flames and in seven minutes, everything was gone. For what? A rat?". . .

As comparable stories sweep the country, more and more landowners learn the lesson. Make sure there is nothing on your land that might attract wildlife or rare species. It will merely bring oppressive attention from federal bureaucrats. Once landowners took pride in private stewardship. Developing and maintaining wide hedgerows around their fields might cost them some income in forgone crops; however, it provided wildlife they could enjoy, bird songs at dawn and dusk, and maybe some quail or pheasant to shoot in the fall. But increasingly that has become too risky. Monoculture crops and sterilization of marginal land are more rational. The "shoot, shovel, and shut up syndrome" inspired by the ESA is rapidly becoming the norm.

The extensive media coverage of the devastating California fires, and especially the television coverage, has finally shown urban America what the ESA is all about. They've seen that it harms people and wildlife. Hopefully, at last America is ready to rewrite the ESA so that it accommodates both man *and* beast. At least two critical amendments are necessary.

Mending the Act

First, the listing process needs to be addressed to reduce frivolous and unscientific listings and listings patently directed at achieving ends other than helping species, i.e., using the Act to halt development or resource use. Currently, the major way for a species to be taken off the Endangered Species List is "original data error," which means that it was subsequently discovered that the species never should have been placed on the list to begin with. Such erroneous listings and corrective delistings represent the squandering of millions of dollars and years of effort in attempting to save something that never required saving. Meanwhile, truly endangered species go down the drain. An amendment should mandate anonymous peer review of listing proposals so an objective panel of scientists would review each listing proposal and comment on whether the scientific evidence is complete and accurate.

Secondly, the Act must be amended to state that no actions mandated by it should take private property or cause a loss of economic use or value of private property without full and just compensation. When the government undertakes actions for the public good, the Constitution requires that private property owners be compensated. Not only will this return justice and equity to the ESA, but with landowners once again secure in their property rights they will no longer have to fear having wildlife habitat or wildlife on their lands and we can return to the era of private stewardship and conservation that predated the Act.

VIEWPOINT 2

"The reasonable regulation of land use is both constitutional and in the public interest."

Private Property Regulation Is Necessary to Save Endangered Species

Bruce Babbitt

Bruce Babbitt is the secretary of the interior in the Clinton administration. In the following viewpoint, Babbitt supports the constitutionality of the federal government's regulation of private property in order to save endangered species. He argues against measures designed to compensate landowners who have lost income due to federal protection of endangered species on their property. Such measures would require the government to pay prohibitively large amounts of money to property owners who have been breaking federal environmental laws, Babbitt maintains. Federal environmental agencies can use creative solutions and management changes rather than financial compensation to address the concerns of private property owners affected by the Endangered Species Act, he concludes.

As you read, consider the following questions:

1. In what ways have local governments regulated private land usage, according to the author?
2. According to Babbitt, how may legal cases alleging a taking under the Endangered Species Act have actually been filed?

Bruce Babbitt, "The Endangered Species Act and Takings: A Call for Innovations Within the Terms of the Act," a speech delivered to the Society of Environmental Journalists, Duke University, October 22, 1993.

The Endangered Species Act is the most innovative, wide-reaching and successful environmental law that has been passed in the past quarter century. I can cite case after case: the resurgence of the American alligator, the fact that the skies are once again graced by many bald eagles, and that the peregrine falcon is moving from near extinction to the threshold of de-listing. The opponents of the Endangered Species Act know these facts. So they have come at us in a different direction, advocating a new and radical concept—that any government action lowering the value of someone's property creates a right to be compensated by the U.S. Treasury. For example, H.R. 1388, styled as the "Just Compensation Act of 1993," would require federal agencies to compensate property owners "for any diminution in value" caused by any regulatory action taken under environmental laws, including, right at the top, the Endangered Species Act [H.R. 1388 died in the U.S. House of Representatives in 1994].

Let's examine the implications of this proposed raid on the public treasury. The Kesterson National Wildlife Refuge in California is one of the great migratory bird stops on the Pacific flyway. But a few years ago, the waterfowl were dying, and they were deformed at birth. It turned out to be selenium poisoning running off into the refuge from nearby farm irrigation wastewater. Under the Endangered Species Act, I tell the farmers: Clean up the pollution, or we'll sue you. But under this new proposal, I am undeniably causing "a diminution in value" of a property right—it will cost those farmers money to clean up. They'll comply, but then they'll send me the bill! The old legal maxim, "Make the polluters pay," would be replaced by a new legal rule: "It pays to pollute; the government will reimburse your costs."

Setting a Precedent

If H.R. 1388 were to pass, who knows what would be next. When the Environmental Protection Agency bans pesticides that have been found to cause cancer, the chemical companies that have incurred losses will send the bill to our government. When the Food and Drug Administration takes a breast implant off the market, the companies will send the bill to the FDA. Where do you stop? There will never be enough money to pay everyone for every government regulation.

In fact, the American courts have always recognized that the reasonable regulation of land use is both constitutional and in the public interest. Every day, in every part of the country, city councils and county zoning commissions routinely make land use decisions that affect land value, adding to some land while subtracting from others. Consider my hometown, Flagstaff, Arizona, located high in the ponderosa forests of northern Arizona. In the early 1980s, the City Council used its zoning powers to pass a law making it a crime to cut down a pine tree on private land inside city limits unless it is to make space for improvements authorized by the planning and building codes. The residents of Flagstaff supported the law. People like to live there because you can smell the perfume of the pine forest in the air, and you can see an extraordinary horizon in every direction. Admittedly, they have subtracted from the freedom of a landowner who wants to saw down every pine tree on his lot, but they are protecting the overall image of their town. Palmdale, California, has an ordinance prohibiting the removal of Joshua trees from private land. And there are myriad other examples.

It is undeniable that the Endangered Species Act limits the ability of some landowners in some places to do anything they want—to raze the forest, to bulldoze the habitat, to dry up a stream which contains an endangered species. And after listening to our opponents, I figured there must be some cases of egregious abuse. We went to the Court of Claims, where there are hundreds of "takings" cases of all kinds being filed in waves of protest and you know what we found? In the 20 years of this Act, when we've listed some 800 species, there has not been a single case alleging a taking under the Endangered Species Act. The fact that the Fish and Wildlife Service has never come close to a constitutional taking does not end the matter. The government has a higher obligation to the citizens who elected us than simply staying out of court.

Strategies for Protection

The first step is to ask: Are there public lands that can be the core of the protection scheme? Have you read about the California spotted owl? No. You know why? By good fortune 99 percent of their habitat is on public land up in the Sierra Nevada and northern California. Whenever we can rely on public land, we do.

Another approach is flat-out mitigation. Take the example of the desert tortoise in the Great Basin of Nevada, California and Arizona. Several years ago, the city fathers of Las Vegas, a boom town if there ever was one, discovered that all subdivision land had already been taken by the desert tortoise. It got there first. So, we worked out a plan that said to the developers, "as you bulldoze tortoise habitat for subdivisions, we will levy a surcharge on each lot, like a surcharge for water, sewage and roads, that we will use further out to buy up private lands that are inholdings in the public domain to set up tortoise reserves." We can rearrange the protection landscape.

In some cases management changes work. The endangered red cockaded woodpecker hangs out all around the South. But it's a more manageable bird than the spotted owl. It has to live in an old growth pine tree, but it's not at all picky about its neighbors. You'll find them living on golf courses, backyards—all you have to do is make sure it's got good shelter, and the supermarket is not too far away. It also has excellent family values. It's a monogamous critter. And the young ones stick around to help raise the next generation. Really an admirable bird, worthy of our protection. So we got the Georgia-Pacific company to develop a plan for their biologists to travel ahead of the logging crews, identify the old growth woodpecker trees and keep a modest habitat circle around that tree. This procedure will impact maybe one percent of their timber land.

No One Owns the Land

Human beings who live less than a century claim land that has been there since the dawn of time as "ours." We maintain the right to "develop" this land, to behave as if the only time frame that mattered were our own lifespan. . . .

But in the end, we don't own nature any more than we own the birds at the feeder. Or the owls in the forest. Whatever fine points the lawyers for the timber industry can draw in a court, nature draws other laws. We can't save the owl and cut down the forests, any more than we can destroy our own habitat and survive.

Ellen Goodman, *Liberal Opinion Week*, May 1, 1995.

In Texas, we had one that looked like a train wreck in the making. Along the Balcones escarpment in central Texas, there are marvelous springs that give rise to the headwaters of rivers that run down to the Gulf of Mexico. As fate would have it, large spring-fed pools hold a few uncharismatic critters living at great depths, including the blind Texas salamander. But pumping outside San Antonio was lowering the groundwater table, which was drying up the pools. And it looked like a disaster. The politicians were going wild: "It's going to be people or the blind Texas salamander—a future for Texas or turn it over to cave-dwelling invertebrates." We took a close look and found that what Texas really needed was a groundwater management plan. The Texas legislature, in a quite thoughtful process, has now passed its first groundwater management plan in history, which will protect the pools and make San Antonio's water supply more secure.

Why do you keep reading stories about hardships? The tough case is a small landowner on a strategic piece of property. When a species is listed, there is a freeze across all of its habitat for two to three years while we construct a habitat conservation plan, which will later free up the land. The *Reader's Digest* story is always about a small landowner caught in this regulatory freeze. So I've told the Fish and Wildlife Service that we want to reduce the inconvenience to a bare minimum on a guy who's saying, "I'd like to clear an acre to build a house for my mother-in-law." We can do that with new concepts, like a transition habitat plan.

The Endangered Species Act is not the problem. The problem is that the people who have administered it haven't creatively explored the range of possibilities in this wonderfully expansive law. I believe that we can preserve the incomparable biodiversity of the American landscape and we can accommodate a reasonable expectation of any landowner. All we have to do is work together.

VIEWPOINT 3

"You can't wipe out the livelihoods of tens of thousands of people just to accommodate the spotted owl."

Saving the Northern Spotted Owl Costs Loggers Their Jobs

Randy Fitzgerald

In the following viewpoint, Randy Fitzgerald argues that restricting logging in the Pacific Northwest old-growth forest—ostensibly to save the habitat of the northern spotted owl—has resulted in the unemployment of thousands of loggers and millworkers. Whole communities have suffered from the logging ban as businesses close and families move away, Fitzgerald contends. Instead of putting old-growth forests completely off-limits to loggers, he suggests, government officials should allow new techniques that will permit loggers to keep their jobs while providing sufficient habitat for spotted owls to survive. Fitzgerald is a reporter and editor for *Reader's Digest*.

As you read, consider the following questions:
1. How has the spotted owl been used as a tool to stop logging in old-growth forests, in Fitzgerald's opinion?
2. According to the author, how will U.S. consumers be affected by the old-growth restrictions?
3. In what ways is New Forestry an improvement over clear-cutting, in Fitzgerald's view?

Randy Fitzgerald, "The Great Spotted Owl War." Reprinted with permission from the November 1992 *Reader's Digest*. Copyright ©1992 by The Reader's Digest Assn., Inc.

One afternoon in November 1991, Donald Walker, Jr., got a four-page letter from an attorney for an environmental group calling itself the Forest Conservation Council. The organization threatened to sue, seeking heavy fines and imprisonment, if Walker cut down a single tree on his 200 acres of Oregon timberland.

Walker, his wife Kay and two daughters live in central Oregon on land that has been in the family for three generations. Since being laid off from his lumber-mill job in 1989, Walker had cut a few trees each year for income to help support the family.

Barely able to control his mounting anger, Walker called his 77-year-old father, who lives nearby. He had received a similar letter. Then Walker talked to a neighbor, a retired log-truck driver, who cut timber on his land just to pay his property taxes every year. The same threat had been mailed to him.

Kay Walker reacted with disbelief. "Here we are caught up in this owl mess again," she fumed.

The "mess" had begun when environmentalists challenged the Interior Department's decision not to list the northern spotted owl as a threatened or endangered species under the Endangered Species Act. The act permits anyone to sue to enforce provisions protecting a species in peril and its suspected habitat.

As a result of this and related court action, most timber sales on federal forest land in the Pacific Northwest have been halted, throwing thousands of loggers and mill employees out of work.

Now the act is being used against private landowners. Besides the Walkers, about 190 other landowners in Oregon have received legal threats from the same environmental group. Most are small private landowners or modest local logging companies. "We can't afford to fight this in court," Walker says. "I'm out of work, and last year our property taxes nearly doubled. Our tree farm is the last hope we have to survive."

The hardships visited on logging families by the spotted-owl controversy will eventually touch all Americans through higher prices for wood products. But these problems could have been avoided—and still can be—if environmentalists, timber owners and federal officials would compromise.

The Real Agenda

In 1987, a Massachusetts group called Greenworld petitioned the U.S. Fish and Wildlife Service to list the northern spotted owl as an endangered species. After a review, the FWS ruled that the owl was not in danger of extinction. In retaliation, 22 environmental groups—ranging from the Seattle Audubon Society to the Sierra Club—sued to reverse the decision.

A number of these groups had another agenda—to outlaw logging in old-growth forests throughout much of the Northwest—

and were using the owl as a tool. "The northern spotted owl is the wildlife species of choice to act as a surrogate for old-growth forest protection," explained Andy Stahl, staff forester for the Sierra Club Legal Defense Fund, at a 1988 law clinic for other environmentalists. "Thank goodness the spotted owl evolved in the Pacific Northwest," he joked, "for if it hadn't, we'd have to genetically engineer it."

Old-growth forests are often defined as stands of trees at least 200 years of age that have never been exposed to cutting. There are nine million acres of old-growth forest on federal lands in Oregon, California and Washington. Of this, some six million acres—enough to form a three-mile-wide band of trees from New York to Seattle—are already off-limits to logging, preserved mostly in national parks and federal wilderness areas.

A Logger's Story

In November 1991, I received a letter from an outfit called the Forest Conservation Council telling me that if I cut any more timber on our [family tree farm] it would sue me for violating the Endangered Species Act, which protects spotted owls, and makes it a crime to tamper with their habitat.

I have never seen a spotted owl on our place, and I have never met anyone from the Forest Conservation Council. So far as I know, it's never even been on our farm. But I do have a typewritten, single-spaced four-page letter from their lawyer saying that what we have been doing on our tree farm for 60 years is no longer legal.

I might have felt a little bit better about the letter if they had offered to buy the land, or at least pay the taxes, which we have also been doing for 60 years. But they didn't and I guess I'm not surprised. From what I've read about these people, they don't believe in private property rights.

Donald Walker, Jr., *The Wall Street Journal*, May 15, 1992.

So the fight came down to the remaining three million acres, which were being cut at the rate of some 60,000 acres a year. By the time this old growth was harvested, foresters for the Northwest Forestry Association argued, a like amount of acreage in other forests would have matured into old growth. Environmental groups countered that the spotted owl would be extinct by then because it can't survive in sufficient numbers in younger forests.

Responding to the environmentalists' petition, U.S. District Judge Thomas Zilly ordered the FWS to take a second look. Then, U.S. District Judge William Dwyer stopped most Pacific

Northwest timber sales on U.S. Forest Service land. And in June 1992, U.S. District Judge Helen Frye banned old-growth timber sales on most of the Bureau of Land Management's Pacific Northwest land.

In June 1990, the FWS reversed course and listed the owl as threatened, after a committee representing four federal agencies concluded that the owl population was declining. The estimated 2000 owl pairs still alive, the committee decided, were dependent primarily on the old-growth timberland. The FWS has since proposed a critical spotted-owl habitat in the three states, with suggested sizes ranging from 11.6 million to 6.9 million acres.

Later academic studies have challenged the government's conclusions. A timber-industry group, the American Forest Resource Alliance, summarized 15 studies by forest experts at major universities and discovered that, as more land is surveyed, the known owl population continues to increase. Even the FWS's 1992 projections show 3500 known pairs, nearly twice the number federal bureaucrats first estimated.

Furthermore, the Alliance contends that the owls do not require old-growth forest; they can adapt to younger forests. Northern spotted owls thrive in Boise Cascade's 50,000-acre forest near Yakima, Washington, which has been harvested and regrown repeatedly. The same situation exists on 70,000 acres of Weyerhaeuser timberland near Eugene, Oregon.

Despite the evidence, the wheels of government and the federal courts have been set in motion to protect the owl. The result has been havoc for people.

"Alaska Widows"

Nestled in a picturesque river valley at the foot of Oregon's Cascade Mountains, the town of Oakridge calls itself the tree-planting capital of the world. Its 3400 residents are surrounded by the Willamette National Forest, which teems with elk and deer, bear and cougar.

After timber-sale restrictions began to take effect, logging companies started laying off workers, and truckers who had hauled the wood were idled. Mill workers who had been making as much as $17 an hour found that the few jobs available were sacking groceries or pumping gas at minimum wage, and even those soon disappeared.

Local unemployment shot up to 25 percent. For-sale signs sprouted like mushrooms. Businesses began to go bankrupt—first the variety store and the animal-feed store, then three gas stations, two clothing stores, several restaurants and the town's only movie theater.

To survive financially, several dozen Oakridge men sought employment in the only section of the West Coast still hiring loggers—Alaska. Separated from their families ten months at a time, they live on rafts in a region accessible only by floatplane or boat. Left behind in Oakridge are the wives, who call themselves Alaska Widows.

Cheryl Osborne, who has three children, rises each morning at 5:30 to cook breakfast in the restaurant she and her husband opened in the building next to their house. The house is up for sale, and the restaurant is barely making it. In the afternoons Osborne works as a bookkeeper for a small logging company that's just making ends meet.

Linda Cutsforth hasn't been able to find full-time work since she lost her mill job after 25 years of employment. She has seen the strain take its toll on timber families. "Loggers look like whipped dogs," she says. "My husband feels like he's sentenced to prison in Alaska."

Jill Silvey works at the local elementary school, where she has seen the economic casualties up close. One fourth-grade boy lived in a tent on the river with his family after they lost their home. Several children from another family live in a campground and arrive hungry at school each day.

In timber towns across the Pacific Northwest, families and entire communities that had once been close-knit are disintegrating. Loggers in towns with names like Happy Camp and Sweet Home, who had taken pride in their self-sufficiency and hard work, now feel abandoned and betrayed.

Adding Up Costs

"Environmentalists predicted in 1990 that only 2300 jobs would be affected in the three states," remembers Chris West, vice president of the Northwest Forestry Association.

Earlier this year, the FWS projected the loss of 32,100 jobs. As compiled by timber-industry groups and labor unions, the ultimate figure, taking privately owned woodlands into account, may exceed 100,000.

Ripple effects have begun to reach consumers nationwide. Pacific Northwest states supply more than one-third of all the softwood lumber and plywood produced in America. In 1991 the volume of wood withdrawn from harvest because of owl restrictions was enough to construct 270,000 new homes. The scarcity drove up lumber prices at least 30 percent, adding more than $3000 to the cost of building a $150,000 home.

If the restrictions on cutting continue, most economists expect a further sharp rise in timber prices. For every 20-percent increase in wood costs, up to 65,000 American families are priced out of houses they could have afforded previously. Prices will also rise on paper products and furniture.

Short-term relief would be available if the Forest Service could salvage wood presently rotting on the ground. Major storms, for example, have blown down 195 million board feet of timber in Oregon's spotted-owl habitat, enough to keep more than 1300 people employed for up to a year and provide enough timber to construct some 16,000 American homes.

But environmental groups have blocked the Forest Service from salvaging the wood—this despite the government and industry contention that not all of the blow-downs are essential to the owl's habitat.

Striking a Balance

"There have certainly been forest abuses," admits Cheryl Osborne, herself a former member of the Audubon Society. Clear-cutting, for example, leaves large, ugly bald spots sprinkled with the charred remains of stumps and debris. Congress, too, is to blame, having directed the overcutting of trees in the national forests to increase federal revenues. "But you can't wipe out the livelihoods of tens of thousands of people just to accommodate the spotted owl," declares Osborne. "Why can't there be a balance?"

There can be, if loggers use a range of techniques known as New Forestry. At Collins Pine Company's 91,500-acre forest in northeastern California, no clear-cutting is permitted. Old-growth-forest trees such as ponderosa pine are mingled with other species of new growth. Most trees killed by insects, burned or blown down are weeded out, but vigorous ones are left to help replenish the forest with seed.

At Collins, more trees are always growing than are being cut. The result is a thriving wildlife population, including bald eagles, ospreys and California spotted owls.

"We should be able to manage forests for spotted owls," says wildlife biologist Larry Irwin. "We know of hundreds of cases where owl habitat was created by accident as a result of management practices. Surely, then, we can do it by design."

Most environmental groups are skeptical of New Forestry: it still means cutting trees. Many timber companies resist it, claiming it is a less efficient way to harvest fewer trees. A growing number of foresters and wildlife biologists, however, are accepting New Forestry as a bridge to cross the deep chasm that separates most environmental groups from most timber growers.

Spotted owls and logging are not incompatible—and Congress must take this controversy away from the courts and carve out a compromise that serves the national interest. "The reign of terror against private landowners must end," says Donald Walker, Jr. "Loggers need their jobs back, the Alaska Widows need their husbands, and the nation needs the renewable resource that this group of hard-working Americans provides."

VIEWPOINT 4

"Many of the jobs would soon disappear, owl or no owl."

The Northern Spotted Owl Is Not Responsible for Loggers' Unemployment

Alexander Cockburn and Timothy Egan

In the following two-part viewpoint, Alexander Cockburn and Timothy Egan contend that the restriction of logging in old-growth forests in order to save the northern spotted owl is not the cause of unemployment among loggers. In Part I, originally published in 1992, Cockburn argues that the timber industry's own overharvesting had led to the layoffs and cutbacks that company officials attempted to blame on the owl. Cockburn is a columnist for the *Nation*. In Part II, Egan maintains that, by 1994, many former workers in the Oregon timber industry had found new and fulfilling careers, reflected in Oregon's low unemployment rate. Egan is the Seattle bureau chief for the *New York Times*.

As you read, consider the following questions:

1. According to Cockburn, why is the corporate timber industry desperate to get access to federally managed forests?
2. What types of jobs are former timber workers training for, according to Egan?
3. According to Bill Morrisette, as quoted by Egan, why are companies such as Sony Corporation attracted to Oregon?

Alexander Cockburn, "Owls vs. Kids on T-Shirts? Don't Buy It," *Los Angeles Times*, January 13, 1992. Reprinted with permission. Timothy Egan, "Oregon, Foiling Forecasters, Thrives as It Protects Owls," *The New York Times*, October 11, 1994. Copyright ©1994 The New York Times Company. Reprinted by permission.

I

Once again the spotted owl stands accused of threatening workers' livelihoods—indeed, the well-being of the timber economy of the whole Pacific Northwest.

In January 1992, the federal Fish and Wildlife Service designated 6.9 million acres of forest as critical to the owl's survival, noting that at current rates of harvest, the remaining old growth will have disappeared in 15 to 30 years.

Already the timber industry and its allies have fastened on the jobs issue. The Fish and Wildlife Service predicted that preservation of spotted owl habitat would cost about 33,000 jobs. The industry is already throwing around numbers like 80,000. Soon loggers will be sporting company-supplied T-shirts with owl-versus-hungry-kids logos; the right-wing talk shows will be abuzz with the owl recipe du jour.

Fraud

The campaign is a fraud, and the internal corporate memos are there to prove it. These memos, which have been circulating in the environmental community for two or three months, surfaced publicly in the January 5, 1992, edition of the *Santa Rosa (Calif.) Press Democrat* in an article by that newspaper's timber correspondent, Mike Geniella.

The memos, written by a 16-year senior employee of Louisiana-Pacific (L-P), one of the timber giants of the Northwest, and by an L-P consultant who was formerly director of the California Department of Forestry, show that responsible people inside the industry have been saying exactly the same things as environmentalists.

On March 19, 1990, just as the Redwood Summer campaign was warming up, with Louisiana-Pacific's flacks denouncing Earth First activists as ignorant troublemakers, Bob Morris, resource chief for the company's western division, was writing to L-P chief Harry Merlo: "It now appears we are currently harvesting at a rate nearly double our sustained yield (i.e. growth volume)." Morris described this as a resource philosophy akin to "liquidation."

At the end of 1990, on December 18, Jerry Partain, a former state forestry department director regarded as an industry advocate, summed up a consulting assignment for L-P in a private memo to Joe Wheeler, the company's western division manager: "It is now clear to me that the environmental activists, the Department of Forestry, your contract loggers and your foresters are all correct when they say your present harvest rate cannot be continued for long and that the company must either reduce that rate significantly now or make deep and drastic cuts just a few years from now" (an assessment in which the state forestry board recently and publicly concurred).

Partain stated that L-P's logging of small-diameter trees was creating a large hole in its timber inventory "that cannot be recovered for decades." By August 1991, not long before Morris was fired (he's now back with the company), he wrote bitterly to Merlo: "The management of L-P owes its stockholders, long-time employees, their families and the dependent communities a better legacy than is now unfolding."

These smoking-gun memos put the spotted owl frenzy in perspective. Corporate timber is desperate to get full access to publicly owned forests because it has recklessly degraded its private resource.

Owls Are Not Responsible

As specious as the industry's pretensions to responsible resource management is the notion that the owl's future must be balanced against 33,000 jobs. Employment in the timber industry along the North Coast has dropped for decades, long before the owl became an issue. The Fish and Wildlife Service said many of the jobs would soon disappear, owl or no owl. L-P's western division work force dropped from 2,700 in 1988 to 1,700 today because of mill closures and its export of unfinished redwood logs to its new plant in Mexico.

The spotted owl is an "indicator species," meaning its impending extinction signals a much wider threat in terms of loss of habitat, biodiversity and the maintenance of commercially important gene pools of plants and trees. A threat to the owl is a threat to much else, including the tourist economy of the Northwest. Who, except real estate developers, wants to stroll through clear-cut old growth?

The destruction of old growth is already very far advanced, with the big timber corporations left undisturbed in their rampages for decades. The best thing would be for the government to declare eminent domain and give the huge private timber empires to local loggers and have them harvest the timber responsibly on a sustained yield basis. Better them than the moguls of lumber, issuing orders to pillage from their corporate headquarters in Atlanta (Georgia Pacific), Portland, Ore., (Louisiana Pacific) and Houston (Maxxam, owner of Pacific Lumber).

II

By now, the timber communities of Oregon were supposed to be ghost towns. There was going to be an epidemic of foreclosures, a recession so crippling it would mean "we'll be up to our neck in owls, and every millworker will be out of a job," as President George Bush predicted in 1992 while campaigning in the Northwest.

Politicians in both parties agreed. The villain was the northern spotted owl, an endangered bird fond of the same ancient national forests desired by loggers. When restrictions on logging were ordered in 1991 to protect the bird, Michael Burrill spoke for many of his fellow Oregon timber mill owners when he said, "They just created Appalachia in the Northwest."

But economic calamity has never looked so good. Three years into a drastic curtailment of logging in Federal forests, Oregon, the top timber-producing state, posted its lowest unemployment rate in a generation, just over 5 percent.

What was billed as an agonizing choice of jobs versus owls has proved to be neither, thus far. Oregon is still the nation's timber basket, producing more than 5 billion board feet a year. (Ten thousand board feet are used to build the average house.) But instead of using 300-year-old trees from public land to make 2-by-4's, mills are relying on wood from tree farms, most of them belonging to private landowners. And the mills are getting more out of the timber, using parts that used to be discarded.

Technical Jobs

Between 1989 and 1994, Oregon lost 15,000 jobs in forest products. But it gained nearly 20,000 jobs in high technology, with companies like Hewlett-Packard, which makes computer parts, expanding considerably in the state. By early 1995, for the first time in history, high technology surpassed timber as the leading source of jobs in the Beaver State. And timber workers are being retrained for some of those jobs, particularly in manufacturing.

Instead of spectral monuments to the spotted owl, many parts of the state have reached what economists call full employment—a jobless level of about 5 percent that the experts say will not cause inflation and where people are usually unemployed by choice. And there are signs of impending labor shortages, according to state economists. In 1993 alone, the state's growing economy has added nearly 100,000 jobs—the exact amount the timber industry said would be lost with the restrictions.

Even the most timber-dependent counties in southern Oregon report rising property values and a net increase in jobs.

But some in the timber industry say the crash is yet to come. Many mills are using trees that should not have been cut because they are too small, said Chris West, a spokesman for the Northwest Forestry Association, an industry group based in Portland.

"The small woodlot market blossomed more than anyone expected," Mr. West said. "But it's going to be short-term."

Asked about the job-loss figure of 100,000, Mr. West said, "We don't think the hammer has hit yet."

Retraining

As for the loggers and millworkers who have already lost their jobs, most of them did not become minimum-wage hamburger flippers, as predicted. At Lane Community College in Springfield, the nation's largest center for retraining displaced woodworkers, nearly 9 of every 10 people going through the program have found new jobs, at an average wage of $9.02 an hour, about $1 an hour less than the average timber industry wage. They are becoming auto mechanics, accountants, cabinetmakers and health care workers.

"So many people say this is the best thing to ever happen to them," said Jeff Wilson, a former millworker from the town of Mapleton, who is just finishing his retraining program and plans to become a community service worker. "I was brain-dead at the mill, never thought I'd do anything else. Now, it's like the world has opened up."

Preserve Old-Growth Forests and Jobs

The state of second-growth forests has been almost forgotten in the focus on the so-called spotted owl controversy. Protection of what is left of our ancient forests is a national and even a global need. We need lumber and jobs, too, but these needs can be met from already cut-over lands in national and state forests and in privately owned timber lands. Now in second and third growth, these forests can provide jobs, lumber, protection of soil and water, protection of some forest plant and animal species, and continued scenic and recreation values if properly managed.

Virginia Warner Brodine, *People's Weekly World*, April 17, 1993.

The big question on retraining, one that President Clinton brought half the Cabinet here to discuss at the timber summit in the spring of 1993, was what a timber worker could be retrained to do. It turned out to be a simple answer, said Patti Lake, who runs the retraining program.

"I'm so sick of the Paul Bunyan stereotype about these people," Ms. Lake said. "They come to us because they know there are better jobs than burger-flipping. They're just people who graduated from high school and went to work in the mill or the woods. Now, they're becoming the accountant who does my taxes or the mechanic who fixes my car."

To be sure, there are pockets of poverty in the smaller, more remote timber towns of Oregon. The aid package promised by President Clinton, $1.2 billion over five years, has only begun to trickle in. Under the President's plan, the timber cut in national forests will be about one-fourth of what it was in the 1980's.

Places like Sweet Home and Oakridge have lost Main Street businesses as the mills have closed. Auctions of heavy equipment used to haul and mill giant trees are common.

High Employment Rate

But no county in Oregon has an unemployment rate higher than 7.8 percent, and in some rural counties, the rate is about 2 percent, compared with the national rate of 5.9 percent. Also, few people seem to be leaving. During the last period of timber layoffs, from 1981 to 1987, Oregon lost population. In 1993, the population grew by 40,000 people.

And as the number of logging jobs has fallen, the average wage here has risen. In 1988, the peak year for timber cutting, wage levels here were 88 percent of the national average. In 1994, they were 93 percent.

In 1991, Representative Bob Smith, a Republican from the eastern part of the state, said the logging restrictions "will take us to the bottom of a black hole." And that year Representative Peter A. DeFazio, a Democrat who represents the biggest timber-producing district in the nation, in south and western Oregon, sketched a picture of widespread devastation.

But here in Lane County, in Mr. DeFazio's district, the unemployment rate is 4.8 percent. Mr. DeFazio, who still predicts some economic downturn, said he has been pleasantly surprised by some of the positive developments.

"I met a guy my age, a timber worker who was being retrained to become a nurse," he recalled. "A nurse! He said, 'Yeah, it's the fulfillment of a lifetime dream.'"

Springfield, the blue-collar neighbor of Eugene, across the Willamette River, has landed a new Sony Corporation factory, where compact discs will be manufactured. It may employ 1,500 people within five years, at salaries that will start at better than $30,000 a year.

"It wasn't blind, dumb luck that helped us land Sony," said Mayor Bill Morrisette of Springfield. The company wanted a pristine place on the river, he said.

Using some money from President Clinton's forest recovery package, the town offered Sony $8 million in tax abatements and incentives. In return, Sony promised to pay people at least 10 percent above the national average. Today, the factory is rising on farmland just miles from the woods that have been shut down to logging to protect the spotted owl.

Quality of Life

"Owls versus jobs was just plain false," Mr. Morrisette said. "What we've got here is quality of life. And as long as we don't screw that up, we'll always be able to attract people and business."

And even though numerous timber mills have closed in Springfield because they could no longer get the big trees, newer, leaner operations like Springfield Forest Products are hiring. The Springfield mill, which was shut in 1989, was retooled to use small-dimension wood from tree farms. When it was owned by Georgia-Pacific, the mill relied on old-growth timber from national forests.

The mill now employs 450 people.

"A lot of people were afraid of change," said Scott Slaughter, the personnel manager, a third-generation timber worker. "But I see a real future here."

Ed Whitelaw, a professor of economics at the University of Oregon in Eugene, was one of a handful of economists who predicted that job losses would be minimal and that Oregon, because of its attractive scenery and low property costs, would thrive.

"These 100,000-job-loss figures were just fallacious; they came out of a political agenda," Dr. Whitelaw said. "Yet when I would say this, I was dismissed as an Earth-Firster or something."

Mr. Burrill, who owns a timber mill in Medford, Oregon, was asked about his statement that saving the spotted owl would create Appalachia in the Northwest.

"We've had an awful lot of new industry, and that's surprised me," he said. He said people moving to southern Oregon from California are not all retirees, as the stereotype has it.

"They are bringing jobs with them," he said. "Turns out there's a hell of a lot going on."

VIEWPOINT 5

"These carnivores may displace ranchers—which is what many environmentalists hope."

Wolf Reintroduction Threatens Ranchers' Livelihoods

Alston Chase and *Human Events*

In the United States, due largely to an extermination policy during the early part of the twentieth century, wolves are virtually extinct in the wild in all states except Alaska. The following two-part viewpoint argues that the reintroduction of these endangered wolves into their former habitats will hurt the livestock industry. In Part I, nationally syndicated environmental columnist Alston Chase contends that reintroduced wolf packs are likely to kill livestock, as well as other endangered animals and domestic pets. In Part II, the national conservative weekly *Human Events* describes negative reactions toward wolf reintroduction in the western states and in North Carolina. Residents of these areas fear that wolves will quickly become a public nuisance and may even attack children, *Human Events* reports.

As you read, consider the following questions:

1. In Chase's opinion, what three scenarios may lead to wolves' preying on domesticated livestock rather than on wild deer or elk?
2. Under what conditions would Chase approve the return of wolves to Yellowstone National Park?
3. According to *Human Events*, how has the plan to reintroduce the endangered red wolf into North Carolina gone awry?

Alston Chase, "Game of Chance with the Wolf Pack," *The Washington Times*, January 27, 1995. Reprinted by permission of Alston Chase and Creators Syndicate. "States Revolt Against Wolf Program," *Human Events*, February 10, 1995. Reprinted by permission.

I

As Western ranchers and Interior Department officials snarl at each other over wolf reintroduction in Yellowstone National Park, the nation seems confronted with yet another environmental cliche. The $7 million federal scheme to plant 15 wolves in the park each year, say the media, pits a regional interest group against the environment—a theme we've heard a zillion times before.

The selfish interests in question this time supposedly belong to stockmen, who have sued to stop reintroduction and who doggedly insist that any canine that could eat "Little Red Ridinghood's" grandmother and then wear her clothes is perfectly capable of scarfing leg-of-lamb once in a while.

Environmentalists dismiss this as a fairy tale spread by rednecks who lack tolerance for cross-dressing mammals. Wolves are necessary for "ecological balance," they insist, and piggish people who fear fanged intruders huffing and puffing and blowing doors down shouldn't live in straw houses. Meanwhile, the feds insist they are merely enforcing the Endangered Species Act, which mysteriously requires wolf "recovery" in Yellowstone but not in, say, Scarsdale, New York, or Washington's Rock Creek Park.

But this fight isn't what it seems. It is not a local issue. It is not about welfare cowboys and has nothing to do with "ecological balance." Rather, it raises questions about the fairness of a policy that will ultimately affect everyone.

Tinkering with the Environment

Wolf "recovery" is another step toward fulfilling the environmentalist agenda of "returning" the continent to "presettlement conditions." But rather than resembling the past its effects will be unique and unpredictable. Federal "preservation" policy has already created unprecedented conditions that threaten the very values it seeks to enhance. Leaving wilderness alone ignores how Native Americans shaped the landscape through burning and produces forests that are older and less biologically diverse than when Columbus landed. Overly protected game populations have reached levels never seen before. These creatures damage vegetation and provide unlimited food sources for mountain lions, whose numbers in many regions are greater than any time in natural history.

These same misguided schemes transformed Yellowstone. Elk, surpassing original numbers by 50,000 or more, have consumed habitat important to other animals such as white-tail and mule deer, bighorn sheep and beaver. Introducing wolves into this unusual mix will produce results no one can anticipate. Authorities hope the new arrivals will eat elk and won't bother cattle.

But that may not happen. In the last century, these canines revealed a decided taste for livestock. And in Minnesota (which has 2,000 "threatened" wolves of its own) predation on domestic animals is surging.

Stiff Regulations

Under the terms of the U.S. Fish and Wildlife Services wolf-reintroduction program, ranchers can shoot only wolves caught in the act of killing or devouring their livestock. Since cattle and sheep are often spread out across hundreds of acres, however, many ranchers stand a better chance of winning the lottery than of nabbing a wolf *in flagrante*. Even if he should be so lucky, the rancher had better hope he actually owns the ill-fated cow or sheep. Only the owner of livestock under attack may use lethal force against the reintroduced wolves....

As stiff as the terms of the wolf-reintroduction program are, they will get stiffer if environmentalists have their way. The Sierra Club Legal Defense Fund has sued to bring the wolves under the long arm of the Endangered Species Act, which would prohibit ranchers from shooting, harming, or harassing the wolves in any way, even if they catch them brunching on their livestock. Worse, federal wildlife managers could fence off thousands of acres now used for grazing or recreation if they deemed those activities a threat to the wolves' health, shutting down ranches and throwing hundreds of people out of work.

Valerie Richardson, *National Review*, March 20, 1995.

Rather, "reintroduction" is an ecological crap-shoot. Several scenarios are possible:
(1) Wolves may prefer to eat deer. But deer, whose park habitat has been colonized by elk, stay mostly in surrounding farmland. Ergo, wolves will loiter in hay pastures and develop a fondness for hamburger.
(2) Wolves may "key" on Yellowstone's bighorn sheep. Since these creatures are already stressed by competition with elk, they will be wiped out quickly. Then, wolves may switch to the bighorn's domestic cousins.
(3) Wolves may take elk. Since this food is abundant, canine numbers skyrocket. Elk will decrease and wolves multiply until legions of hungry canines invade celebrity enclaves like Jackson, Wyoming, and Aspen, Colorado, taking occasional Labrador retrievers.

In any case, wolves will continue their march across the continent. Alaska has 6,000, and packs are multiplying in North

Carolina and Arizona. A recovery program is under way in Idaho. Minnesota's wolves are increasing 3 percent to 5 percent a year and have reached Hinckley, not far from St. Paul. A southward migration from Canada is under way in the Rockies.

Ranchers vs. Wolves

Over time, these carnivores may displace ranchers—which is what many environmentalists hope. Thus government is forcing rural landowners to subsidize the atavistic pleasures of others in the name of "ecology." The feds won't pay for damages done by its animals. And although the Defenders of Wildlife, a conservation group, promises to compensate for losses to cattle (and not to sheep), its funds might not suffice. Minnesota, which disburses up to $40,000 a year for stock losses, has found that amount insufficient. Most kills go unreimbursed, and pet owners never receive compensation.

The issue, therefore, is fairness. While wolves should indeed return to Yellowstone, they must not become Trojan horses for primitivists who wish to dismantle civilization. Justice requires the animals be confined to the park and that realistic control and compensation systems be established.

But this probably won't happen until wolves take up residence on the yuppie spreads proliferating around Yellowstone. When carnivores begin eating upscale dogs and scaring children, the politics of this species will get really interesting.

II

As the Interior Department's controversial wolf "reintroduction" program continues to wreak great havoc on local communities, several states have called for defiance of this new wrinkle in the administration of the federal Endangered Species Act (ESA)—and even repeal of the ESA itself. Beyond the serious challenge to one of the cornerstones of the environmental agenda, the raucous grassroots defiance is yet another palpable expression of the growing state rebellion against federal edicts.

The reintroduction of wolves—animals not seen in substantial numbers in most of the United States since the 1920s—was launched after the Interior Department determined that "species recovery" was a requisite part of the 1971 ESA. Over great local opposition, the federal authorities started a "red wolf" program in a region of eastern North Carolina in 1987 and, in February 1995, a "gray wolf" program in Yellowstone Park.

Countering local citizen and rancher objections to the program in both regions, federal officials repeatedly insisted that each program was required by law and assured citizens that surrounding communities would be virtually unaffected by the wolves.

But in February 1995, concerns about the new Yellowstone

program were almost immediately confirmed. Just a few days after the release of 19 Canadian gray wolves into the Yellowstone wilderness in Idaho and Wyoming, one was shot and killed some 60 miles from where it was released after attacking and devouring a new-born calf.

State Resistance

Immediately expressing its anger, the Idaho legislature voted to refuse any further state cooperation with the federal measure. And at congressional hearings on the $7-million initiative late in January 1995, freshmen Representatives Helen Chenoweth (R.-Idaho) and Barbara Cubin (R.-Wyo.) berated the wolf-release program as misguided and an assault on states' rights.

Confronting Secretary of the Interior Bruce Babbitt, Chenoweth said, "I strongly believe, Mr. Secretary, that not only have your wolves trespassed onto the lands of Idaho, but you have trespassed onto the Constitution of the United States."

But while Interior Department officials were trying to downplay the initial problems with the new Yellowstone wolf program, problems continued to intensify with the red wolf plan in Hyde County, North Carolina, where, after seven years, the state is now mounting open defiance to the program.

Contrary to department assurances that the red wolves would remain relatively small in size, docile and confined to the state's wilderness areas, they have become a public nuisance and hazard.

Following a quick spread throughout three counties, many of the wolves are now twice their predicted size and are depleting the state's big game. They frequently wander through residential areas, have been responsible for killing a number of livestock, prized hunting dogs and pets, and have staged two near attacks on humans.

But far from responding to the state's pleas to discontinue its wolf plan apparently gone awry, federal authorities have kept up the program and may even be increasing their enforcement efforts. In February 1995, a Hyde County resident, who accidentally trapped and killed a red wolf on his own property, but then immediately contacted the Fish and Wildlife Service, was potentially facing federal charges under the ESA.

Keeping Kids Indoors

A staffer to Rep. Walter Jones (R.-N.C.), the congressman who represents most of the affected areas and who is seriously considering seeking an end to the $350,000 annual congressional appropriation for the program, said, "This has become an extremely serious problem in our district. In some areas, people see them quite frequently and they are understandably reluctant to allow their pets and children go outside. There is a lot of anger and fear about this program."

Fred Bonner, a local North Carolina biologist who was called to testify in the congressional hearings in January 1995, said, "What has people most upset down here is that the federal government has come in and effectively taken away both their property rights and their community's way of life by forcing this terrible plan on them."

Angry about the worsening conditions, the North Carolina legislature at the beginning of January 1995 quickly passed, with broad bipartisan support, a bill authorizing citizens in the three affected counties to repel the wolves. With federal officials likely to challenge this North Carolina statute as a breach of the Endangered Species Act, the issue is probably headed for a showdown in federal court.

"We don't know what will happen if this thing goes to court," commented Bonner. "But the people who worked on this statute made it very clear that these are non-migratory animals, that they believe this is purely a state issue and that they can do this."

Troy Mayo, a Hyde County commissioner who drummed up support for the measure, said, "I do feel that [the courts] are going to continue to cram this thing down our throats whether we want it or not. The only way we'll get relief and get our community back is if the people of the United States begin to get disturbed about this."

VIEWPOINT 6

"Wolves aren't the cause of the changes occurring in the West."

Wolf Reintroduction Does Not Threaten Ranchers' Livelihoods

Renée Askins

Renée Askins is the executive director of the Wolf Fund, a Wyoming organization that works for the reintroduction of endangered wolves into Yellowstone National Park. In the following viewpoint, which is excerpted from Askins's January 1995 testimony to the U.S. House Committee on Resources, Askins argues that the reintroduction of wolves will not greatly affect ranchers. Contrary to popular belief, wolves would kill far less livestock than die each year from natural causes or from attacks by dogs, she maintains. Wolves are an important part of the western habitat, Askins concludes, and therefore should be reintroduced.

As you read, consider the following questions:

1. What basic operational cost has the livestock industry transferred to the general public, in the author's view?
2. According to Askins, why did wolves disappear from the Yellowstone region?
3. In what ways have both ranchers and environmentalists oversimplified the issue of wolf reintroduction, in the author's opinion?

From Renée Askins's testimony before U.S. House of Representatives Committee on Energy and Natural Resources, January 1995.

If I were a rancher I probably would not want wolves returned to the West. If I faced the conditions that ranchers face in the West—falling stock prices, rising taxes, prolonged drought, and a nation that is eating less beef and wearing more synthetics—I would not want to add wolves to my woes. If I were a rancher in Montana, Idaho, or Wyoming in 1995, watching my neighbors give up and my way of life fade away, I would be afraid and I would be angry. I would want to blame something, to fight something, even kill something.

The wolf is an ideal target: it is tangible, it is blamable, and it is real. Or is it? When ranchers talk about wolves they say, "You know, it's not the wolves we're worried about, it's what the wolves represent; it's not what they'll do, it's what they mean." Wolves mean changes. Wolves mean challenges to the old ways of doing things. Wolves mean loss of control. Wolves aren't the cause of the changes occurring in the West any more than the rooster's crow is the cause of the sun's rising, but they have become the means by which ranchers can voice their concern about what's happening around them.

Ranchers deserve our compassion and our concern. Whether the threat of wolves is imagined or actual, the ranchers' fear and anger are real. I honor that. However, it is my job, as a scientist and an advocate, to distinguish fact from fiction and purpose from perception.

Wolves and Livestock

Ranchers claim that wolves will devastate the livestock industry in the West. Yet all the science, the studies, the experts, and the facts show that wolves kill far less than 1 percent of the livestock available to them. According to the *Bozeman (Montana) Chronicle*, even if federal specialists have wildly underestimated the number of cows and sheep that wolves would kill in the Yellowstone and central Idaho areas, the actual total would be much smaller than the number that die each year in the state of Montana alone because of storms, dogs, and ovine ineptitude. In fact, the number of wolf-caused sheep deaths would have to be almost thirty times higher than predicted before it matched the number of Montana sheep that starved to death in 1993 because they rolled over onto their backs and were unable to get up.

In effect, the livestock industry has successfully transferred to the general public one of its most basic operational costs: prevention of predator losses. If you raise Christmas trees, part of the cost and risk of doing business is losing a few trees to gypsy moths and ice storms; inherent in the cost of ranching, particularly on public lands, should be the cost and risk of losing livestock to predators. Instead, every year 36 million tax dollars go to kill native predators on our public lands so that private industry can make a profit.

It is important to remember that wolves are missing from the Yellowstone region only because we eliminated them. They did not vanish from the area in response to loss of prey or lack of habitat; they did not die out as a result of disease or natural catastrophe. We systematically, intentionally, consciously killed every wolf we could find (and we found all of them). And we didn't just remove wolves that killed livestock. The wars against predators at the turn of the century weren't about ridding ourselves of a nuisance; they were about the principle of dominance, and the wolf, the symbol of wild, untamable nature, was the object of conquest. We didn't just want to control wolves, we wanted to conquer them. So we didn't just kill wolves, we tortured them. We lassoed them and tore them apart by their limbs; we wired their jaws shut and left them to starve; we doused them with gasoline and ignited them.

Minimal Impact

The Farm Bureau has argued that wolf reintroduction would put ranchers out of business. . . . Data supporting its arguments simply do not exist. Ranchers and livestock owners have been coexisting with the wolf for many years in Alberta, British Columbia, Minnesota, western Montana, Wisconsin, and North Carolina with minimal conflicts. Wolf predation in all cases averages less than one-tenth of one percent.

In all three existing recovery efforts, and in Yellowstone and central Idaho, extraordinary efforts have been taken to ensure that livestock damage and economic impact are minimal.

Robert M. Ferris, *National Debate*, March/April 1995.

Opponents of wolf reintroduction assume that because there are no wolves, there should be no wolves, and over the last two decades they have effectively framed the reintroduction debate around that assumption. They have promoted the idea that the return of wolves is somehow radical or extreme, some sort of environmental luxury, some romantic nonsense that only urbanites and rich Easterners advocate at the expense of the poor, beleaguered Western livestock industry. (In fact, surveys of the residents of Wyoming, Montana, and Idaho show that Westerners support the reintroduction.) The industry's cry of economic loss has eclipsed the costs to the general public of not having wolves. In the West we now live in a "wolf-free" environment. Or is it "wolf-deprived"? Who has gained and who has lost? How do we assign a value to the importance of a predator in the ecosystem? How do we determine the cost of removing one note from a Mozart symphony, one sentence from a Tolstoy novel, or one brush stroke from a Rembrandt? Having wolves in Yellowstone is not a luxury but a right. We should not have to pay for clean air or water, nor should we believe that they are somehow a luxury. Similarly, we have a right to a full complement of wildlife on our public lands.

Oversimplification

Because of the passion and politics involved, it is easy to oversimplify this debate. Just as unrealistic as the ranchers' scare tactics are the claims by certain environmentalists that wolves are sweet and docile animals; that the wolf is the ultimate symbol of harmony; and that everything noble, wise, and courageous is somehow embodied in this one creature. According to this view, ranchers, hunters, and industry are the bane of the environment, and saving the wolf, no matter what the cost, will be our redemption. Both environmentalists and ranchers have used exaggerated rhetoric to alarm constituents. Both advocates and opponents of reintroduction have tried to use wolves as the "line in the sand" that divides the old West from the new. Both sides want us to see this issue as a distillation of all endangered-species conflicts, as a simple question of either/or: don't touch a tree vs. clear-cut all trees; no wolves vs. fully protected, untouchable wolves; unrestricted grazing vs. no grazing. The tragedy is that while the armies fight their wars, the rest of America stands by, confused, uncertain, and unaware that something they care about might be at stake.

The truth, as always, lies somewhere in the middle. Wolves are not killing machines that deserve hideous deaths; neither are they cuddly creatures needing tender, righteous protection. Wolves survive by killing; they have an extraordinary and complex social system; they are smart, strong, and, at the core, consummate predators. Restoring wolves will not rescue us from our economic or ecological troubles, but neither will their presence contribute to them. Some ranchers may indeed lose sheep or cattle to wolves. Some [hunting] outfitters may find fewer elk for their clients to shoot. Yet neither ranchers nor outfitters will face economic doom due to the presence of wolves.

Emotions, not facts, have controlled the Yellowstone wolf debate. Wolves have never been just wolves: the wolf is the devil's keeper, the slayer of innocent girls, the nurturer of abandoned children, the sacred hunter, the ghostly creature of myth and legend. In short, wolves are symbolic; Yellowstone is symbolic; restoring wolves to Yellowstone is a deeply and profoundly symbolic act.

Symbolic Force

We are a culture of symbols. It is not surprising that ranchers and environmentalists use the symbolic force of wolves to debate painful changes. We distill ideologies into incidents, processes into people; we use symbols to help us order and make sense of an increasingly complex world. The Yellowstone wolf-recovery debate is fundamentally an expression of a culture in transition; it is the struggle that accompanies old assumptions clashing against the new. The story of this conflict is the story of how we view ourselves in relation to animals, whether we can replace the assumption of "dominion" that has been so destructive to us and the natural world with a worldview that recognizes that we live in a state of reciprocity with the birds and the beasts—that we are not only the product of nature but also part of it. Our attitudes toward wolves and our treatment of them cut to the very marrow of how we view our relationship to the natural world. Is wolf recovery loss or enrichment? Relinquishing or sharing? Wolves mean something to everyone. But in the end, wolves are only wolves. The real issue is one of making room, and there is still a little room in the West—room for hunters, for environmentalists, for ranchers, and for wolves.

7 VIEWPOINT

"Some 7,000 'indicator species'... have been used... to wreak havoc with property owners."

Wetland Regulations Are Unfair to Property Owners

William F. Jasper

Wetlands are an important habitat for many endangered or threatened species, as well as other nonthreatened species that rely on wetlands for their survival. What constitutes a wetland, however, is an issue of fierce debate. In the following viewpoint, William F. Jasper argues that the definition of "wetland" has become too broad and subjective. Jasper asserts that current wetland regulations have resulted not only in loss of income for many private property owners and developers, but even in unjust prison sentences for some. Environmentalists and government officials are employing these all-encompassing wetland regulations to gain jurisdiction over immense amounts of privately owned land, he contends. Jasper is the senior editor of the *New American*, a conservative biweekly, and the author of *Global Tyranny... Step by Step*.

As you read, consider the following questions:

1. How did the 1989 redefinition of wetlands affect Bill Ellen's waterfowl sanctuary project, in Jasper's view?
2. What are the three factors by which wetlands are now delineated, according to the author?
3. According to Jay H. Lehr, quoted by Jasper, what is the aim of the new definition of wetlands?

From William F. Jasper, "Eco-Villains?" *The New American*, February 8, 1993. Reprinted with permission.

His prison ID number is 27445-037. His name is William B. Ellen. On November 30, 1992, after losing a three-year legal battle that pitted him against the United States Justice Department, the FBI, the Army Corps of Engineers, the Soil and Conservation Service, and the Environmental Protection Agency (EPA), Bill Ellen entered the Petersburg Correctional Camp, a federal prison at Petersburg, Virginia.

In the closing days of 1992, thousands of petitions, letters, phone calls, and telegrams poured into the White House appealing to President George Bush to pardon Ellen. The pardon effort was organized by the Fairness to Land Owners Committee (FLOC) and Alliance for America, organizations networking with property owners groups nationwide to defend property rights. But the United States government had expended enormous resources—manpower, tax dollars, and political capital—to put this dangerous felon behind bars and President Bush was not about to release this menace to society at the behest of some motley, misguided letter-writing campaign.

A Dangerous Criminal

What kind of heinous criminal is Ellen that authorities would decide to throw the full weight and power of the federal government into the effort to bring him down? A top lieutenant to Mafiosa don John Gotti or Colombian drug kingpin Pablo Escobar, you guess? Not even close. An Iraqi-paid assassin, mass murderer, carjacker, or kiddy porn ringleader? Wrong again.

Ellen is not the kind of man one usually associates with *America's Most Wanted*. But to the powerful environmental lobby that apparently controls the Justice Department's prosecution priority list, the Bill Ellens of this world are vicious "environmental criminals" who rank in the "Public Enemy Number 1" category.

The 47-year-old eco-villain was convicted in January 1991 on five of six counts of violating Section 404 of the federal Clean Water Act by destroying wetlands without a federal permit. Truly dastardly acts deserving of the harshest punishment, say the green crusaders. Yes, filling a wetland is very serious stuff these days. It conjures up ugly images of greedy capitalists bulldozing the Everglades, dumping tons of mercury into pristine riparian ecosystems, or paving over the last aquatic habitat of the snowy egret and the furbish lousewort.

But that isn't what Ellen was doing. Indeed, Bill Ellen was *creating* wetlands. Yes, Bill Ellen was constructing a waterfowl sanctuary, complete with duck ponds, marshes, and wetland vegetation—on what was previously dry land. He was planning "to create duck heaven." For this "crime" he is now serving time in the federal slammer—as a wetland *destroyer*. Confusing? As we shall see, there is very little in the absurdly convoluted and muddled federal wetlands policies that isn't confusing.

Bill Ellen's troubles started in 1987 when he accepted a job creating wetlands on the 3,200-acre Maryland estate of wealthy New York commodities trader Paul Tudor Jones. Jones had in mind to create, as the centerpiece of his estate, a 103-acre wildlife sanctuary that would attract and support geese, ducks, and other wildlife. Ellen, a conservationist and marine engineer, was hired to supervise construction of the waterfowl habitat on uplands that were so dry water had to be sprayed on the soil as a dust suppressant when work crews began moving dirt around. Ellen consulted frequently with local, state, and federal agencies, including the Army Corps of Engineers, the Soil Conservation Service, the Maryland Department of Natural Resources, and the Dorchester County Planning and Zoning Board. He obtained over two dozen permits and hired environmental consultants to complete ecological surveys of the property to make sure no wetlands were filled.

Reprinted by permission of Jerry Barnett/*Indianapolis News*.

In 1989 the Bush Administration redefined wetlands. Overnight the total "wetland" area in Dorchester County jumped from 84,000 acres to 259,000 acres. Virtually the entire county had been declared a wetland. The Army Corps sent Jones a cease-and-desist order for all work at the estate. On March 3, 1989, Army Corps official Alex Dolgas and Soil and Conservation Service wetlands expert Jim Brewer visited the Jones estate and ordered all work shut down.

Ellen immediately shut down all work at the sanctuary except for construction on the management complex, a three-acre site where a couple of houses and a kennel were under construction. He pointed out to Dolgas and Brewer that Brewer had inspected the area only a month previously and had agreed that it did not contain wetlands. Work on the complex was already behind schedule and Ellen was facing penalties from architects and contractors if he delayed further. Ellen told the two regulators he could have another wetlands survey done by an independent consultant within 48 hours and would shut down work on the complex if the survey showed that the area did indeed include wetlands. But he did not want to shut down and break contractual obligations, only to find that the federal government's fickle paper shufflers had goofed again. (As the judge presiding over Ellen's trial noted: "The fact that a government employee says a permit is required does not necessarily make it so.")

According to Ellen, Dolgas wouldn't accept the offer for a new survey and "got in a huff, jumped back in his truck and left." Ellen went to a phone and called the landscape architect who was the overall supervisor of Jones' project. After talking to the architect and reconsidering the matter, Ellen decided to comply with the Corps order. He told the foreman to halt the project. It was too late. Two truckloads of clean fill dirt had already been dumped on the dry ground of the work site. That was enough to cost Ellen his freedom. . . .

Definitions

What constitutes a wetland? There is no political or scientific consensus today on what is a wetland, says attorney Mark L. Pollot, author of *Grand Theft and Petit Larceny: Property Rights in America.* "The term 'wetlands' itself is not a scientific term and only began appearing recently in scientific literature," Pollot said. "But it is being defined and applied in outrageous ways by environmentalists, politicians, judges, and bureaucrats to deny property owners their rights guaranteed by the Constitution."

Under the Carter Administration, wetlands were defined as areas flooded or saturated with ground water often enough that, under normal circumstances, they would support "vegetation typically adapted for life in saturated soil conditions." The definition emphasized that wetlands were limited "to only aquatic areas"—i.e. bogs, swamps, and marshes. That alone was a good-sized bite of new federal power.

But that didn't satisfy for long. Since then definitions have changed drastically and have extended Washington's jurisdiction over vast areas of dry land. Wetlands are now delineated by three highly elastic technical factors: hydrology (the wetness of the soil), the presence of "hydric" soil (usually soil with a peat, muck, or mineral base), and the presence of hydrophytic vegetation (plant varieties that tolerate standing water or waterlogged soil).

The land grabbers realized that the wetland vegetation parameters presented limitless possibilities. The Corps of Engineers, which is in charge of issuing permits for all activities on wetlands—with the EPA holding veto power—developed guidelines using a classification system of five plant types to help distinguish swamp vegetation from that found on dry land. It is those guidelines, which have evolved into a list of some 7,000 "indicator species," that have been used by the Corps and the EPA to wreak havoc with property owners. The vegetation guidelines were incorporated into the *Federal Manual for Identifying and Delineating Jurisdictional Wetlands*. This has provided the regulatory socialists with defining parameters sufficiently broad and arbitrary to claim jurisdictional control over not only every mud puddle in the country, but over completely dry land that could not be considered wetlands by the furthest stretch of the imagination. Eureka, bureaucrat heaven!

A Tyrannical Policy

The EPA insisted on including "facultative vegetation"—plant species that appear in uplands as often as wetlands—as a wetland-defining parameter. The presence of Kentucky bluegrass, poison ivy, impatiens, ash trees, dogwood trees, or any of hundreds of other facultative species now provides the federal wetlands gestapo with sufficient cause to ruin your day—and your life. According to Robert Pierce, a former Corps of Engineers regulator who helped write the guidelines for Section 404, the vegetation parameters have been transmogrified into completely nonsensical and tyrannical policy. One of the most common facultative plants, he notes, is the red maple tree which can grow in standing water—or on the top of a mountain! "What is being called a wetland," says Pierce, "is not functionally different from uplands."

The original 1989 *Wetland Manual*, says Dr. Jay H. Lehr, a world-renowned water scientist, "condemns as much as 300 million acres of mostly private property to a useless future in spite of the fact that it may appear high and dry to the 'untrained' eye."

Dr. Lehr warns that "the rank-and-file citizenry, which has often been willing to stand up for the rights of swamp critters, is being ambushed by the broad new definition of a wetland being fostered by environmental zealots. It is aimed far more at limiting the rights of the individual in favor of the higher causes of 'society' than actually giving two hoots for the native flora and fauna of our strange new dry, and often barren, 'wetlands.'". . .

A Dedicated Environmentalist

Bill Ellen is hardly your typical eco-villain. A lifelong conservationist and environmentalist, Bill and his wife Bonnie have run a nonprofit animal rehabilitation center called Wildcare on their seven-acre farm in Mathews County, Virginia, since 1986. They have saved over 2,000 injured ducks, geese, hawks, eagles, egrets, deer, and other animals. They were contributors to Greenpeace, the World Wildlife Fund, the National Wildlife Federation, and the Audubon Society. Until she became a mother, Bonnie ran the county's humane society. To support his family and the always growing menagerie of nature's unfortunates, Bill has worked as an environmental consultant. For several years prior he worked as a state wetlands regulator.

His incarceration [Ellen was sentenced to six months in jail, four months of home detention, and one year of supervised release] means both economic and emotional hardship for Bonnie and the couple's two young sons, but she's "sure that we'll make it through." Volunteers from throughout the county are helping with the animals. Friends, family, and total strangers have helped with important moral support. "My husband pays for all our animals' feed and some of it, like the fawn's milk, is very expensive. Without Bill's income it's very tough." Then there's the mortgage and legal expenses, of course.

Convicted for Cleaning Up

But Ellen's punishment is not as severe as that meted out to fellow enviro-criminal John Pozsgai. Pozsgai, a self-employed mechanic, was sentenced to 27 months in prison and a fine of $202,000. His crime: cleaning up thousands of tires and rusting car parts that littered the property he had bought for the purpose of building a new repair shop. After removing the tons of junk, he spread clean fill dirt on part of the site, an activity that state officials told him required no permit. But federal officials said the presence of "skunk cabbage" and "sweet gum trees" made it a wetland. He spent nearly two years in Allenwood Federal Prison for his "crime." Now out of jail, John Pozsgai must still battle the envirocrats to get permission to build on his own land.

"It's destroyed our business, it's been absolutely devastating to our family, financially and emotionally," Pozsgai's daughter, Victoria Pozsgai-Khoury, said of the prolonged struggle. "No American family should have to go through what we have gone through." The experience sorely tried Mr. Pozsgai's faith in the American system of justice, one of the attractions that had led him to flee to this country from Hungary in 1956.

"When you put dirt on top of dry dirt in your own backyard, where the water table is nine feet below the ground, and they throw you in prison, what kind of justice do we have?" Mrs. Pozsgai-Khoury asks. "My family escaped from a communist country where the government can take your land away and came to America where that kind of thing is not supposed to happen. It's as though the government is trying to make it impossible for the private citizen to own property, or to enjoy the freedom and prosperity that comes from owning your own land."

8 VIEWPOINT

"Wetlands continue to be sacrificed for short-term economic goals with little regard for the long-term environmental . . . consequences."

Wetland Regulations Are Fair to Property Owners

Douglas A. Thompson and Thomas G. Yocom

Douglas A. Thompson is chief of the Wetland Protection Section of the Boston office of the U.S. Environmental Protection Agency. Thomas G. Yocom is a wetland program expert at the EPA's San Francisco office. In the following viewpoint, Thompson and Yocom contend that many of the stories regarding wetland-regulation infringements on property owners' rights are exaggerated or fictitious. Moreover, they suggest, the development of privately owned wetlands affects the public good, which validates government regulation of these ecosystems. The protection of wetlands is of greater importance than the construction of such developments as shopping malls, Thompson and Yocom conclude.

As you read, consider the following questions:

1. What types of species rely on wetland ecosystems, according to the authors?
2. What principle for allowing or forbidding wetland development do Thompson and Yocom propose?
3. According to the authors, in what ways does forbidding wetland development prevent public harm?

From Douglas A. Thompson and Thomas G. Yocom, "Uncertain Ground," *Technology Review*, August/September 1993. Reprinted with permission from *Technology Review*, ©1993.

An old woman in Wyoming cannot tend her rose bed because the government accuses her of disturbing protected wetlands. A hard-working Hungarian immigrant spruces up his property, removing tires and other debris; the federal government jails him for filling wetlands. Farmers hesitate to plow and plant their fields for fear of running afoul of federal wetland regulations. Reading stories like these in the national press, one might suppose that tax-supported bureaucrats are prowling the countryside like vigilantes, trampling the rights of innocent property owners. In fact, the parable about the rose garden, which appeared in a *Wall Street Journal* editorial, is fiction. John Pozgai, the Hungarian immigrant, filled nine acres of wetlands in defiance of local, state, and federal regulatory agency warnings and a court order requiring him to stop. And for the past 15 years, most agricultural activities already under way have been exempt from wetlands regulation.

Such horror stories are but one symptom of the intensifying conflict over the future of America's wetlands, a clash rooted in the ambivalence many Americans feel about the public value of environmental protection versus the rights of property owners. This schism is nowhere more divisive than in the area of wetlands regulation, for three-quarters of the nation's wetlands lie in private hands. Ever more vigorous enforcement of wetlands regulation has provoked bitter opposition from both industry and groups championing the rights of property owners. . . .

The 20-year history of wetlands protection reveals a combustible mixture of science and ideology. And the debate, while important in its own right, has important implications for other critical environmental programs, such as the protection of endangered species, in which the rights of property owners may at times conflict with the common good.

The Importance of Wetlands

Known variously as swamps, marshes, bogs, and fens, wetlands form at the interface between land and water. Some are covered or saturated with water throughout the year; others are partially or completely dry for weeks or months. Only the composition of the soil, or the presence of characteristic plants such as cattails, bulrushes, and red maples, may indicate that a given area is wetland.

U.S. wetland ecosystems range from permanently frozen Arctic tundra to languid deep-water swamps in the South; from Dakota prairie potholes and northern quaking bogs to wet meadows in mountain regions. Although saltmarshes and mangrove swamps ribbon portions of the coast, more than 97 percent of the nation's wetlands are freshwater systems. In some areas, such as the tundra of Alaska or the lowlands of Louisiana, wetlands dominate the landscape; elsewhere, the natural limitations of climate or sustained human depredation have made them scarce. Overall, wetlands comprise 6 to 9 percent of the surface area of the lower 48 states.

The dual nature of wetlands—neither entirely land nor entirely water—in part explains the competing views of them within our society. The same area may be an important hydrological and biological component of a watershed and yet be zoned and taxed as residential or commercial property.

Over the last 20 years we have learned much about the consequences of destroying wetlands. These areas are among the most productive ecosystems on earth, vital to the nation's environmental and economic health. Our fish and wildlife populations, including many rare and commercially important species, rely on wetlands for survival. Two-thirds of the species of Atlantic fish and shellfish that humans consume depend upon wetlands for some part of their life cycle, as do nearly half of all federally listed threatened and endangered species. More than 400 of some 800 species of protected migratory birds, for example, rely on wetlands for feeding, breeding, and resting. . . .

Despite their value and relative scarcity, wetlands continue to be sacrificed for short-term economic goals with little regard for the long-term environmental and social consequences. The U.S. Fish and Wildlife Service estimates that only about 95 million acres of wetland—half the nation's original endowment—remain in the lower 48 states. Of these, we are losing more than 250,000 acres each year—about 30 acres every hour. Some states, such as Ohio, Iowa, and California, have lost more than 90 percent of their wetland endowment to agriculture or the construction of highways, reservoirs, industrial complexes, residential developments, and shopping malls.

The Clean Water Act

When the Clean Water Act was passed in 1972, few observers recognized the potential significance of Section 404. This provision regulates the discharge of dredged or fill material into "the waters of the United States," and thus affects a wide array of construction activities in a range of aquatic habitats.

Section 404 was controversial at birth. The U.S. House and Senate agreed that the Environmental Protection Agency (EPA) and the Army Corps of Engineers should each have a role in the new program but differed over which agency should have primacy. In the end, the legislators compromised: the corps would administer the program on a day-to-day basis but would be bound by environmental regulations known as the "404 guidelines," which EPA would develop. EPA also retained the right to veto a corps' decision to grant a permit if the project would have an unacceptably adverse impact on the environment. The agencies share enforcement authority. . . .

Developers and Oil Companies

The prospect of EPA and the corps working harmoniously to protect wetlands was disturbing [to developers]. In the mid-1980s, dam and highway builders, oil companies, and developers saw projects scrutinized more closely, modified more frequently, and, in a small fraction of cases, simply denied. Protests against wetland protection began to intensify.

In 1987, a group called the Wetlands Coalition for Procedural Fairness formed to address what it considered threats to future development. The group eventually metamorphosed into the National Wetlands Coalition, with members such as ARCO, BP America, Exxon, Kerr-McGee, and Unocal. Oil companies take a special interest in restrictions on wetlands development for two reasons: Alaska and Louisiana. To drill for oil in these wetland-rich states, companies must often obtain permits for the construction of access roads, drilling pads, and other facilities.

At the same time that special interests such as oil and construction began organizing against wetland protection, a number of developer-supported groups, such as the Pennsylvania Landowners Association and Maryland's Fairness to Landowners Committee, also entered the fray. Groups in the West, often positioning themselves under the banner of "wise use," began to champion the cause of mining, timber, and grazing interests. *High Country News* reports that People for the West!, which proclaims itself a "grassroots campaign in support of western communities," has a board of directors in which 12 of the 13 members are mining-industry executives.

These groups maintain that the wetland program effectively takes control of private property without just compensation, as required by the Fifth Amendment, and often allege that the federal government prohibits all activities on wetlands. In fact, of the 90,000 activities regulated by the Army Corps of Engineers yearly, 83 percent are authorized by general permits, a form of expedited approval not requiring public notice or hearing. Less than 5 percent of all permit applications are denied, and even permit denial rarely deprives a landowner of all economic uses of a parcel. . . .

A Complex Process

Fundamentally, the protection of wetlands will remain a complex undertaking simply because it has to be carried out on a case-by-case basis, but the case-by-case approach has its limitations. Nearly everyone favors protecting wetlands, yet the environmental value of an individual parcel often seems expendable when weighed against the economic gains derived from jobs, profits, or private use. The result is that few particular proposals appear damaging enough to deny.

We must take a more ecologically sound approach. Just as we evaluate the hazards of smoking and not the ill effects of each single cigarette, we must assess permit decisions in terms of both individual and cumulative impacts. The current principle—under which permits are issued only for the least environmentally damaging, practicable alternative—is effective but difficult to implement. Hundreds of hours and thousands of dollars are spent trying to determine what alternatives can reasonably be considered economically feasible, and the geographic scope within which alternative sites might be located.

Home for Many Species

Almost one-third of the animals on the endangered species list live or depend on wetlands—including manatees, Florida panthers, whooping cranes, American alligators, and the Schaus swallowtail butterfly. Louisiana's swamps provide wintering grounds or resting areas for most of the migratory birds, including two-thirds of North American species of ducks and geese, that trace the Mississippi from its headlands to the open sea. Most of the species prized by commercial and sport fishermen in the Gulf of Mexico depend on coastal wetlands.

Richard Miniter, *Policy Review*, Spring 1991.

We propose a simpler principle: basing permit decisions on whether the public benefit from the type of activity involved requires the destruction of wetlands, not on the alternatives available to the applicant. For instance, rather than focusing on whether a particular developer can find an alternative site for a proposed shopping mall, we should ask whether the public benefit derived from the construction of shopping malls—employment, access to goods and services—can be achieved without destroying wetlands.

Such a principle would categorically deny permits for many types of activities on wetlands—including, we hope, the construction of shopping malls. Other activities, such as the building of marinas, do require access to water and could be weighed on a case-by-case basis. Permits could also be granted for projects that carry important community benefits, such as replacing a substandard bridge or constructing a road through a wetland to straighten a dangerous turn.

This regulatory framework would be more equitable than the

current one, in which the corps may deny permits to developers who have the resources to choose less environmentally damaging alternatives but grant permits to those who do not. Some exceptions would have to be made—for instance, to allow development in a low-value wetland to protect a more environmentally valuable upland site. But in general, this principle would solidify the relative priority of wetlands protection and halt the incremental losses that are steadily eroding our remaining endowment.

Priorities

Finally, the shape of national wetlands policy will be determined by the nation's ability to reconcile its ambivalence about the regulation of activities on privately owned property. A 1992 Times-Mirror poll found a majority of respondents felt that wetland protection should take priority over property concerns. Nonetheless, a significant portion of the public remains troubled by what it sees as growing government infringement upon the rights of property owners. These citizens argue that wetland regulations and even zoning requirements force landowners to sacrifice for the public good. Just as landowners must be compensated for property seized by eminent domain [the right of the government to take private land for public use], the argument goes, so should these losses be compensated, even though no physical taking of property occurs.

So far, the courts have not looked favorably on these claims. One recent decision, *Lucas v. South Carolina Coastal Commission*, indicates that the current U.S. Supreme Court will generally find a compensable "taking" has occurred in the uncommon circumstance where a regulation has proscribed all economically viable uses of a parcel. A permit denial may prevent some uses on the entire property or all uses on some part of the property but rarely denies all uses on the entire tract. Moreover, the court left open the possibility that even a total deprivation of use would not be compensated. Litigation on this issue will undoubtedly continue through the remainder of the 1990s.

Politicians may be willing to go further than the courts in forcing the government to pay for restrictions on private property use. The 1992 Republican Party platform, for instance, stated: "We will require federal agencies to promptly compensate, from their own budgets, for any taking of private property, including the denial of use." If compelled to pay off every applicant whose permit is denied, government agencies might find wetland protection just too expensive.

We believe denials of wetland fill permits do not represent seizures of private property but rather are decisions designed to prevent public harm: flood risks, degradation of surface and groundwater, and damage to commercial and sport fish and wildlife stocks. The government may be able to insulate itself from some challenges by arguing that restrictions that prevent a nuisance cannot constitute a taking.

Placing Limits

Property rights have always been tempered by limits on the extent and nature of development. Holding title to a property does not invest the owner with the right to create a nuisance or put the land to a noxious use. The government has no obligation, in essence, to pay its citizens not to pollute the environment or destroy valuable public resources. . . .

Having lost over half the nation's wetlands, will we adequately protect those that remain? Or will we, on the altar of "balance," allow many small decisions made in individual self-interest to lead to the eventual sacrifice of a priceless natural resource?

Periodical Bibliography

The following articles have been selected to supplement the diverse views presented in this chapter.

Carolee Boyles-Sprenkel	"Wetlands in Chaos," *American Forests*, July/August 1992. Available from 1516 P St. NW, Washington, DC 20005.
Alston Chase	"Spotted Owl of a Different Feather," *The Washington Times*, February 15, 1993. Available from 3600 New York Ave. NE, Washington, DC 20002.
Gregg Easterbrook	"The Birds," *The New Republic*, March 28, 1994.
Timothy Egan	"Ranchers Balk at U.S. Plans to Return Wolf to the West," *The New York Times*, December 11, 1994.
Eric T. Freyfogle	"Owning the Wolf," *Dissent*, Fall 1994.
William F. Jasper	"A Morass of Wetlands Myths," *The New American*, March 8, 1993. Available from PO Box 8040, Appleton, WI 54913.
Wallace Kaufman	"The Cost of 'Saving': You Take It, You Pay for It," *American Forests*, November/December 1993.
Brad Knickerbocker	"Private Property vs. Protection of Species: Two Tales of 'Taking,'" *The Christian Science Monitor*, March 7, 1995. Available from One Norway St., Boston, MA 02115.
Amy Linn	"Guess Who's Loping In for Dinner?" *Outside*, October 1994. Available from 400 Market St., Sante Fe, NM 87501.
Charles C. Mann and Mark L. Plummer	"Empowering Species," *The Atlantic Monthly*, February 1995.
William Perry Pendley	"Do Grizzlies Have the Same Rights as Humans?" *Environment*, Fall 1993.
Ana Radelat	"Who Says Saving the Earth Costs Jobs?" *Public Citizen*, May/June 1992.
James Owen Rice	"Where Many an Owl Is Spotted," *National Review*, March 2, 1992.
John C. Sawhill	"We Can Have Growth and *Still* Save Endangered Species," *USA Today*, July 1991.
Jeff A. Taylor	"Species Argument," *Reason*, January 1994.

CHAPTER 4

What Should Be U.S. Policy on Endangered Species in Other Countries?

Endangered Species

Chapter Preface

In Zimbabwe and Namibia, wildlife officials have been taking chainsaws to their rhinoceros wards—in order to save the animals from poachers. Rhinoceros horns are in great demand in Asian countries as ingredients in traditional medicines and in Yemen for use as dagger handles. The Convention on International Trade in Endangered Species (CITES), of which the United States is a prominent and influential member, banned international commercial trade in rhinoceros products in 1977. The United States has also established trade sanctions against countries such as Taiwan that have not abided by the terms of the CITES ban. Nevertheless, illegal poaching and black-market trade have flourished, with rhino horns fetching up to $20,000 a pound.

Poachers kill the rhino to remove its horn, a practice that has contributed significantly to the rapid reduction of African rhinos: In 1970, Africa had 65,000 rhinos; by 1995, less than 3,000 remained alive. The rhino's horn can be removed without killing the animal, however, and officials have begun to saw off the horns in hope of discouraging poachers. Unfortunately, this tactic has not always been successful. As scientists Joel Berger and Carol Cunningham report in *Science*, in Zimbabwe "a population of 100 white rhinos, with at least 80 dehorned, was reduced [by poachers] to less than 5 animals in 18 months." Wildlife officials believe that poachers may be killing the dehorned rhinos for the small stub of the sawed-off horn. For poachers, especially those from poor Third World countries, the income generated by this illegal activity has proven irresistible, despite the possible ecological costs of driving a species to extinction.

Several African nations would prefer that CITES lift the ban on the rhino horn trade, as well as similar bans on products made from other endangered or threatened species such as elephants and tigers. Many resent the United States for pressuring their nations to ban the lucrative wildlife trade and point out that the United States has also hunted many of its animals to extinction or endangerment. American journalist Raymond Bonner, who lived in Kenya for several years, maintains that the African nations should "do what they decide is best for themselves" concerning the international wildlife trade. Others, such as John Hoyt, president of Humane Society International, argue that developing countries should learn from America's mistakes and protect their endangered animals from extinction. How large a role the United States should play in these countries' conservation of endangered species is among the issue examined in the following chapter.

VIEWPOINT 1

"The time has come to demonstrate to the world that the people of the United States will not support . . . commercial whaling."

The United States Should Condemn Commercial Whaling

Patricia A. Forkan

Since 1986, the International Whaling Commission (IWC) has banned commercial whaling—a measure opposed by Norway, Japan, Russia, and Iceland. In the following viewpoint, Patricia A. Forkan urges the U.S. government to take a leading role in maintaining the international moratorium on whaling. Forkan particularly opposes the efforts of Japan and Norway to reinstitute the hunting of minke whales, a species not classified as endangered but one whose numbers she believes may be much smaller than current estimates report. In addition, she calls for U.S. support for the proposed Antarctic whale sanctuary (the sanctuary was officially created on May 26, 1994). Forkan is the executive vice president of the U.S. Humane Society and the senior vice president of Humane Society International.

As you read, consider the following questions:
1. What is the final destination of whales killed for scientific research, according to Forkan?
2. In the author's opinion, what is the general sentiment of the U.S. population toward commercial whaling?

Patricia A. Forkan, "Can We Save the Whales?" *The Animals' Agenda*, September/October 1992. Reprinted with permission.

Since 1973 I have been an active participant in the global effort to halt the killing of the oceans' whales. I have participated at every conceivable level at each year's International Whaling Commission (IWC) meeting, from a Non-Governmental Observer (NGO), to being on the U.S. delegation, to my most recent role as chairperson for the body of NGOs represented at the IWC each year. I have seen the killing of whales reduced 100-fold since 1973.

In 1973, more than 45,000 whales were killed; in 1991 less than 400 were killed. While it is my firm belief that the death of even one whale represents a tragic and senseless loss to this planet, nonetheless I have been quite proud of the progress we have achieved in reducing the numbers of whales brutally killed for the profit and pleasure of mankind.

Amazing Creatures

A whale to me is among the most amazing and fascinating creatures to ever live upon this earth. Whether one thinks of the "giant" bowheads, singing humpbacks, the nearly vanished great blue whale, or the small cetaceans: the bottlenose dolphins, the beluga and the narwhal, these majestic and fun-loving animals touch me with their sociability, their intelligence and their complexity. I have always felt a sense of wonder upon the realization that creatures so mammoth in scope, both in their own size and the size of their watery environment, can live so harmoniously in their ocean realm, demonstrating both sagacity and benevolence in their daily existence.

Equally passionate have been the emotions I have felt as I have observed firsthand mankind's ignorant and callous involvement with whales, relentlessly pursuing these creatures to the ends of the globe, and developing more deadly and efficient means to brutally maim and kill these remarkable citizens of the sea. Mankind's greed and stupidity in the senseless destruction of these animals for profit have been the motivating force that has caused me to dedicate significant time and energy toward ending the killing of whales. Believe me, when it comes to killing, human ingenuity knows no bounds.

Whales have been set upon and killed by every grisly method one can imagine—from cold harpoons, to explosive devices, to electric lances, to rifle bullets in their bodies; they have felt the pain and suffered the agony of being perceived by mankind as more valuable dead than alive. For years the IWC has been the international battleground upon which the war over whales has been fought. Despite the frustrations, the heartache, and the anger that has so often accompanied our struggles there, I have for the most part felt hope and optimism that not only were we succeeding by dramatically reducing the number of whales killed each year, but that with the realization of a ban on commercial whaling in 1986, we had finally halted forever the large-scale bloodshed of these magnificent friends.

International Agreement Is Necessary

It took decades of study and debate about whaling—during which time many species were driven to the brink of extinction—for the international community to prohibit this ghastly practice altogether. The best possibility for drawing Norway back into the fold may be to point out that agreements like the whaling moratorium are now crucial in many other areas of environmental and economic policy. If other nations follow Norway's example and opt out of these agreements whenever national preferences are at stake, it will be impossible to forge the kinds of international commitments needed to foster sustainable development.

Langdon Winner, *Technology Review*, November/December 1994.

In 1992 in Scotland, at the 44th annual meeting of the IWC, I was disgusted as I watched my worst nightmare unfold. In a series of maneuvers that illustrated a wanton disregard for the opinion of the citizens of the world, who have unequivocally and overwhelmingly voiced their opposition to the commercial hunting of whales, the IWC moved dramatically closer to approving a resumption of commercial whaling. Perhaps most distressing was the refusal of the U.S. delegation to take even a moderate position in opposition.

Whales in Danger

The battle over commercial whaling has been made in large part utilizing "sustainable use" theories. The question of the ethics of killing whales has for the most part been ignored as inappropriate. Ethical arguments were not considered germane to the discussion in any serious sense. What has been the basis for the arguments over commercial whaling has been the question of how much damage such whaling does to the species of whale being hunted. Most species populations were so devastated by 1986 that world opinion (which leans toward ethical opposition to whaling) and conservation (sustainable harvest) arguments both agreed that unless a ban on all commercial whaling was immediately instituted, all species of whales were in danger of becoming extinct. No one really knows how many members of the various species of whales exist in the waters of the oceans of the world today. However, the Japanese as well as the Norwegians maintain that with respect to the minke whale at least, the numbers were sufficient to reinstitute commercial whaling in 1993. These whales were the last to be ruthlessly slaughtered because their comparatively small size made them less attractive financially. Thus the population was not as decimated.

We may not know how many whales are out there, but we do know how many could be killed. Japan would like quotas in the 5,000–10,000 [range] for their Antarctic bloodletting. Norway wants close to 2,000 in the North Atlantic. Finally, I am sick to death of Norway, Japan, and Iceland manipulating the process for years in an effort to keep their whaling industry alive. Japan killed over 300 minke whales in 1992 alone in the name of science, when everyone is aware the whale meat finds its way to the lucrative meat markets of Tokyo and the other Japanese cities where it resells for upwards of $200 per pound.

Their so-called scientific research has been condemned each year by the IWC as useless. I will not elaborate on all the posturing and long-winded, insulting speeches that spew forth from Norwegian and Icelandic whalers about their sovereign right to harvest whales along their coasts. They say it is "cultural imperialism to impose those foreign values on us."

Clearly all whales are residents of the world's oceans and not to be constrained nor regulated by nationalistic boundaries. Indeed, whales are international citizens and in no way can they be morally regarded as anything less than residents of the planet. Norway's actions in 1992 shocked even the most complacent at the IWC, announcing on opening day that they intended to resume commercial whaling in 1993 no matter what the IWC said. Then they proceeded to rub everyone's nose in it even more by commencing "scientific whaling" for 110 minke whales later that week. What frustrates and infuriates me beyond the whalers themselves is the position the United States has taken in this battle, preferring to compromise rather than lead.

American Sentiment

First one must realize that the vast majority of U.S. citizens clearly and loudly have opposed commercial whaling for several decades. Whale watching and financial support and interest in cetacean strandings and protection are widespread in the United States (*Star Trek IV* was one of the most popular *Star Trek* movies in part due to its message of concern for whales). One must wonder if the U.S. government represented at the IWC by the National Oceanic and Atmospheric Administration (NOAA) Administrator, Dr. John Knauss, has any interest in the opinion of Americans, or if it cares more about maintaining friendly diplomatic ties with allies like Norway and Japan. Consider that while New Zealand has voiced strong opposition to the killing of whales on ethical grounds, and in 1992 Great Britain, Australia and Ireland have moved toward that view, the United States still refuses to address the ethical question, preferring instead to treat it like an unwelcome relative.

Consider that when France proposed in 1992 the establishment of a whale sanctuary around the continent of Antarctica (a move which would effectively halt the Japanese killing of minke whales), the United States would not actively support such a resolution. In fact, aggressive U.S. leadership, according to a majority of IWC participants, would have ensured passage of this revolutionary proposal. Consider that when the IWC scientists proposed a formula on which to base future catch limits (quotas), the United States took a leadership role in getting it adopted. This in spite of the fact that there is strong evidence that quotas could be set on populations well below safe levels. In some cases stocks are at such low levels that they cannot hope to recover to safe levels for one hundred years. Consider that the United States refused to include the issue of humane killing in its list of items necessary to be settled before commercial whaling could resume. I believe the time has come to demonstrate to the world that the people of the United States will not support, condone, accept or ignore the resumption of commercial whaling.

If we as a nation have not evolved to the stage of civilization where the killing of such complex and unique creatures will not be tolerated, then what hope can we hold for a world that is fast disintegrating through destruction of life-giving resources for short-term profits? . . .

The Final Chapter

Clearly, given the pollution of our oceans and the continued over-harvesting of the fish stocks of the seas, the resumption of commercial whaling represents an environmental Pandora's Box. But more significantly, more compelling, is the realization that the United States has within its power the ability to either protect these wonderful creatures forever, or to help hasten Armageddon with respect to the future of the world's whales. As I said earlier, since 1973 I have participated in the battle to protect the world's largest marine mammals.

I have wept and laughed, suffered and rejoiced, but always endured to fight another day. What I see before us now is the final chapter unfolding in our relationship with the world of cetaceans. . . .

I believe we can win for the whales, but the battle is now, the time is today, and the effort must be increased beyond anything that has gone before.

"A sound environmental policy should allow for a sustainable harvest of marine resources, including whales."

The United States Should Accept Commercial Whaling

Mari Skåre

Mari Skåre of Stabekk, Norway, is an attorney and adviser to the Norwegian Ministry of Foreign Affairs. In the following viewpoint, Skåre argues that U.S. opposition to commercial whaling stems from cultural differences between the United States and whale-consuming nations such as Norway. Furthermore, Skåre contends, antiwhaling activists and the American media have exaggerated the intelligence and "human" characteristics of whales, creating an unrealistic image of whales and fostering unwarranted disgust toward whaling nations. Countries that follow sound environmental policies in their whaling methods should be allowed to continue hunting whales without U.S. condemnation, Skåre concludes.

As you read, consider the following questions:

1. Who were the most knowledgeable respondents in the international whaling survey, according to Skåre?
2. According to the author, Norwegian environmental policy is based on which six principles?
3. In Skåre's opinion, how does the average American's exposure to wilderness compare with that of the average Norwegian?

From Mari Skåre, "Whaling: A Sustainable Use of Natural Resources or a Violation of Animal Rights?" *Environment*, September 1994. Reprinted with permission of the Helen Dwight Reid Educational Foundation. Published by Heldref Publications, 1319 18th St. NW, Washington, DC 20036. Copyright ©1994.

To protect whale stocks, an alliance has been built between various environmental and animal rights groups. Advocates of animal rights claim that animals—at least, those with certain characteristics—have an individual right to life, while most environmental activists base their arguments on the danger of furthering the depletion or extinction of certain species and on the importance of preserving ecosystems.

Because of new evidence that limited hunting of minke whales would not jeopardize the species' existence or the ecosystem, Norwegian fishermen resumed minke whale hunting in the northeast Atlantic Ocean in 1993. This decision drew much criticism from the international community. Is the moral outcry against Norway's resumption of whaling a result of skepticism or ignorance of the scientific evidence? Or, does the environmental establishment in the United States and elsewhere no longer support a sustainable harvest of the living resources and instead advocate the individual whale's right to life?

International Attitudes

An international study by Milton Freeman and Stephen Kellert, published in 1992 by the Canadian Survey Circumpolar Institute and Yale University School of Forestry and Environment Studies, surveyed people in Australia, Germany, Japan, Norway, the United Kingdom, and the United States about their attitudes toward whales and whaling. Fifty-seven percent of the U.S. respondents confirmed that they "oppose the hunting of whales under any circumstances," and 55 percent disagreed with the statement, "There is nothing wrong with whaling if it is properly regulated." In Norway, 75 percent of the respondents disagreed with the first statement, and the same percentage agreed with the latter.

The survey also revealed that public knowledge of whale populations was very low. Sixty percent of the U.S. respondents believed there were fewer than 10,000 minke whales left in the world, whereas, in fact, there are approximately 1 million. The Norwegian and Japanese respondents were three to four times more likely than respondents of other nationalities to select a correct answer for the minke whale population. Although none of the respondent groups showed a high level of knowledge, the tendency was clear: The less knowledgeable respondents showed more opposition to whaling.

However, the survey showed "common agreement in all six countries regarding the great importance of (i) protecting whale habitat from pollution and disturbance, (ii) maintaining an ecosystem perspective in whale management, and (iii) basing harvest levels upon the best scientific advice." Except for the Japanese, most respondents in each country identified restrictions on hunting methods and strict international control as a top priority.

Is an ecologically sustainable harvest possible? The approximately 77 species of whales fall into two main categories: baleen whales and toothed whales. The blue whale is a baleen and, as the world's largest animal, may weigh up to 150 tons (equivalent to the weight of 1,900 men weighing 80 kilograms each) and reach a length of 30 meters.

The minke whale, the only whale Norwegians hunt today, is a smaller baleen whale that reaches a length of up to 10 meters. Unlike most baleen whales, the minke whale feeds on fish as well as plankton. (The minke whale apparently got its name from a German named Meinche, who mistook a pod of minke whales for blue whales.) In the North Atlantic, there are three stocks of minke whales: the northeast Atlantic stock, the central Atlantic stock, and the west Greenland Atlantic stock. Norwegians hunt in the northeast Atlantic, and that stock is estimated to include approximately 86,700 animals.

Whale Myths

Opponents of whaling often cite the whales' individual and social abilities. For example, Anthony D'Amato and Sudir K. Chopra stated in the *American Journal of International Law* that

> Whales speak to other whales in a language that appears to include abstruse mathematical poetry. They have also developed interspecies communication with dolphins. Whales are the most specialized of all mammals. They are sentient, they are intelligent, they have their own community, and they can suffer.

And Warner Brothers Family Entertainment claimed in its promotion of the movie *Free Willy* that

> Whales are majestic, gentle, warm-blooded mammals that mate for life, travel in family groups, feel pain, and are incredibly intelligent.

Whale rights activists create an image of a "super whale," incorporating features from various whale species, as well as human attributes. Greenpeace has placed advertisements in newspapers pronouncing the death of the minke whale as the brutal murder of our friend and brother, and the organization has put individual whales up for adoption. Brain surface characteristics and size often are held up as proof of whales' high "intelligence." Around the turn of the century, there was a widespread theory that such characteristics were related to intelligence, but according to M. Klinowska, the theory

> received a severe blow when it was found that the brains of several distinguished people . . . showed no outstanding characteristics whatever and were, in fact, disappointingly ordinary. . . . This was just as well, as elements of such work were

being very badly misused to justify repressive racist, antifeminist and colonial attitudes.

There is no evidence that whales have developed extraordinarily sophisticated ways of organizing communities or communicating in comparison with those of other animals. The strategy of promoting whales as the humans of the ocean, with high intelligence, is probably very good for fundraising, but it may prove counterproductive for animal welfare in general because it directs concern exclusively toward the welfare of animals that have "human-like" characteristics attributed to them. Being considered "human-like," "higher," or "more intelligent" is a poor guide to whether an animal suffers. . . .

Norwegian Policy

Norwegian environmental policies on renewable resources build on the following principles endorsed by the United Nations Conference on Environment and Development held in Rio de Janeiro in 1992:

- *Sustainability:* Depletion of the stocks should be avoided, and only the production surplus should be harvested.
- *Biodiversity:* To maintain global biodiversity, all species must be protected against extinction or decimation.
- *Integration:* Each ecosystem and its resident species should be managed as a unit.
- *Right to exploit natural resources:* All nations and groups should have the right to exploit the natural resources available to them within the above principles.
- *Precautionary principle:* All resource management must be based on the best scientific advice, and uncertainty must result in a cautious level of harvesting.
- *Monitoring and control:* All harvesting must be followed up by monitoring and effective control systems.

Norway is not legally bound by the [International Whaling Commission's (IWC) commercial whaling] moratorium of 1982. However, because of the uncertainty concerning the northeast Atlantic minke whale stock and pressure from the U.S. government and environmental groups, Norway introduced a temporary ban on all whaling in 1987. As the scientific data and the management procedure improved and scientific consensus was achieved, the Norwegian government decided in 1992 to permit commercial harvesting of minke whales in 1993. The result of the 1993 hunt was a catch of 226 animals, 69 of which were caught for research. The quota set by the Norwegian government for the 1994 season was 301 animals, including 110 for research.

The whale meat from the catch was used for human consumption and thus differs from the industrial whaling for oil production previously carried out in the Antarctic. Norway has no desire to resume industrial whaling. The income from minke whale hunting is insignificant for the Norwegian national economy, but, for the coastal communities involved in the hunting, the additional income is of great importance.

Double Standard

While the United States opposes the taking of the abundant minke whales, it allows native Alaskans to hunt bowhead whales, by all accounts a truly endangered species. Sustainable use is fine for the United States, but not for Norwegian whalers who don't vote here; they can be sacrificed to the environmental movement. Imposing restrictions on other countries that are not even tenable at home is shameful and should be stopped. . . .

The United States should take a rational, credible position and support lifting the moratorium.

Michael De Alessi, *The Washington Times*, March 15, 1995.

Norway strongly supports continued research to improve scientific data and hunting methods, which are of concern in Norway as well as in the United States. Significant improvements have been made over the last few decades regarding the equipment and gunners' skills. During the 1993–1994 hunting season, there were inspectors aboard every ship to report on hunting methods and the number of animals caught. Compared to the results from the 1985–1986 season, the whales were killed substantially more quickly and therefore more humanely. . . .

Motives of Environmentalists

The whaling issue raises fundamental environmental questions. What strategy should be followed—the sustainable harvest of natural living resources or protection of the animals' individual rights to life? Throughout the last few decades, an alliance has been built between environmentalists and animal rights advocates to conserve the whale stock. The motivation of the environmentalists has been to preserve or conserve the different whale species and the ecosystems to which they belong. The motivation of animal rights groups has been somewhat different. Philosopher Tom Regan writes:

> That an individual animal is among the last remaining members of a species confers no further right on that animal, and its right not to be harmed must be weighed equitably with the rights of any others who have this right. If, in a prevention situation, we had to choose between saving the last two members of an endangered species or saving another individual

> who belonged to a species that was plentiful, but whose death would be a greater prima facie harm to that individual than the harm that death would be to the two, then the [animal] rights view requires that we save that individual. . . . If the choice were between saving the last thousand or million members of the species to which the two belong, that would make no moral difference.

Regan's philosophy is clear and consistent. He advocates abolition of the use of animals in science, dissolution of commercial animal agriculture (humans should not eat meat), and elimination of commercial and sport hunting and trapping. In the "whale war," however, many opponents of whaling seem to advocate whales' rights because of their "special" characteristics and symbolic value for Western city dwellers. Others defend the right to life of all wild animals (as opposed to human-raised livestock). Why should not a cow qualify for an absolute right to life if a whale holds this right? Could it have something to do with the fact that beef from cattle is an important part of the U.S. diet, while whale meat is not?

Most U.S. citizens risk nothing in the whale war. The United States does not have a whaling industry, and its citizens' need for oil is satisfied by other animal and mineral oils. As [U.S. congressman Gerry E.] Studds so wisely pointed out, the United States is, in fact, making great profits on the whale watching industry.

Promoting "Whalemania"

Greenpeace, which seems to be leading the campaign against commercial whaling, has been invoking both theories. On one hand, the organization attacks Norwegian minke whale hunting by claiming that there is no evidence that the minke whale stock is recovering. On the other hand, its policy promotes certain animals as sacred. Greenpeace's philosophy regarding issues it chooses to address is extraordinarily pragmatic; the group chooses battles it feels that it is in a position to win.

Regarding Greenpeace's campaign to protect seals, Wolfgang Fisher, a former director of the campaign in Germany, stated, "It was necessary for us to be big. Therefore, I found it totally legitimate to use a cute animal with big eyes." So, perhaps Greenpeace's strategy in the whale war is simply to gain as large a profit as possible. As F. Golden put it:

> No doubt about it: whalemania is a growth industry. As fund raisers for environmental organizations have long known, nothing is more likely to tug at the heartstrings—or purse strings—than the sights and sound of whales.

Politicians and businesses have learned to take advantage of the whaling controversy, and so they continue to promote "whalemania." By supporting environmental organizations, companies and politicians have been able to buy themselves a green image and thus gain profits and votes. . . .

U.S. Condemnation

Testifying before the U.S. Congress in 1989, Craig Van Note, vice president of the Monitor Consortium of Conservation, Environmental, and Animal Welfare Organizations, referred to Norwegian scientific whaling as perverted. Brian Davies, leader of the International Fund for Animal Welfare, wrote in a letter to "friends" that whaling in the Faeroe Islands is

> an uncivilized harvest . . . the most brutal you can imagine. Peaceful pilot whales are driven together and tricked towards land with the help of their loyalty to a captured comrade from their family group. And there they are quite simply hacked to death. . . . Civilized peoples have a duty to protect them from these sadists.

It is hard to ignore the arrogance and the moralizing reflected in such statements. Most Norwegians live in close contact with the wilderness, the sea, the mountains, or the forests. For the average Norwegian, the idea promoted by Tom Regan and his followers—that people should not hunt animals in the sea and on land because the individual animal has a right to life—is totally alien. It seems that the moral outcry against whaling is stronger the farther away people live from the wilderness. In the United States, people often experience the wilderness only through their car windows, driving into prepared picnic areas and vista points. The condemnation of Norwegian whaling is surprisingly intense, especially considering that it is expressed by people who live in a society where more than 12,000 people were shot dead in 1992.

Humans are biologically equipped to consume both meats and plants. Only because of this ability to use nature to our advantage have we been able to survive as a species. The question is not whether people should exploit natural resources, but how they will do it. A sound environmental policy should allow for a sustainable harvest of marine resources, including whales.

3 VIEWPOINT

"The [ivory] ban is a conservation success . . . because poaching came to a halt almost overnight."

The International Ivory Ban Should Be Maintained

The Animals' Voice and Andre Carothers

In 1989, the Convention on International Trade in Endangered Species (CITES) declared the African elephant endangered; in 1990, trade in all African elephant products, including ivory, was banned. The following two-part viewpoint argues that the ivory ban has been effective in discouraging ivory poachers and allowing elephant populations to recover. In Part I, the animal rights periodical *The Animals' Voice* maintains that the ivory ban is an appropriate and successful conservation method. In Part II, Washington, D.C.–based journalist Andre Carothers argues that the United States should aid African countries that adhere to the ban. He also urges U.S. sanctions on those Asian countries that purchase ivory and other banned animal products.

As you read, consider the following questions:

1. According to *The Animals' Voice*, what characteristic of elephants makes them especially susceptible to poachers?
2. In Carothers's opinion, the proposition to reopen the African wildlife trade founders on what two points?
3. How does income from wildlife park tourism compare to income from ivory sales, according to Carothers?

"Behold Behemoth: Back from the Brink?" *The Animals' Voice*, vol. 6, no. 4 (1993). Reprinted with permission. Andre Carothers, "Market Solutions," *E: The Environmental Magazine*, January/February 1994. Reprinted with permission from *E: The Environmental Magazine*; Subscription Dept.: PO Box 699, Mt. Morris, IL 61054. (815) 734-1242. Subscriptions are $20/year.

I

In 1988, there were more dead elephants on the plains of Africa than there were live ones. Conservationists estimated that the elephant would be gone in ten years at the current rate of killing. But in 1990, a worldwide ban on ivory went into effect and perhaps the tide has begun to turn. . . .

In some African nations, elephants were being killed at a rate of 80–100 animals a day; 100,000 adult African elephants were being killed every year throughout Africa, another 10,000 calves dying in the process. Dr. Iain Douglas-Hamilton, one of the most respected elephant biologists, estimated that there were 109,000 free-roaming African elephants remaining on the globe—down from 1.2 million, their shocking demise occurring from 1978 to 1988. It was believed that at the rate of poaching, the African elephant would be extinct in less than a decade. And why? Because their ivory tusks were selling for $70 a pound, making the ivory industry a $500-million-dollar-a-year business.

Social Animals

Elephants are highly social animals, traveling in herds of close family units, led by an older cow or two. They are devoted herd animals as well, risking their own lives to protect the lives of each other. It is not uncommon, for example, for poachers to machine-gun an entire herd while some members, under gunfire, attempt to lift their dying companions and families to their feet. Shooting elephants, as it were, was like shooting fish in a barrel.

"When an animal is shot and has its face sawn off with a chainsaw, you cannot just sit back and accept it," says Ian Redmond, one of many mortified human beings watching the elephant massacre from afar. "It's a ghastly sight," adds Curtis Bohlen, a conservationist. "It's something you never forget. It's not only a terrible sight, but if you've ever spent any time with elephants—and I've been fortunate enough to—you realize what special creatures they are."

Elephants are indeed special. Weighing in at four tons and reaching heights of nearly eleven feet at the shoulder, African elephants are the largest living land animals on earth, evolving throughout tens of millions of years, long before their human slayers appeared on the planet. Gifted with a versatile trunk, manipulated by 60,000 muscles, elephants stand alone with exceptional talent. Strong enough to carry or push over trees, the elephant's trunk can also take a morsel of food from an extended human hand with as much flexibility and precision as it is given.

Elephants are highly intelligent, they often bury their dead and commit mercy killings, and they have extremely sensitive emotions and acute awareness. They are known for creating wells in drought-stricken areas, thereby providing the only source of water for many other species of animals. They open up dense forests during feeding, allowing seedlings to get light. "Nature's great masterpiece," writer John Donne called them. And yet other humans reduced them to decaying, faceless corpses.

But in 1990, human beings did something extraordinary: a worldwide ban on the trade in ivory went into effect; Kenyan President Daniel Arap Moi set fire to four years' worth of confiscated illegal ivory—12 tons that might have brought $3 million on the underground market—and he urged other African nations to do the same. The African Wildlife Foundation (AWF) and other organizations made worldwide appeals on behalf of the elephants, urging the human population to support the ban and to cease its consumption of ivory. In the end, the ban passed.

"The ban is a conservation success," says Paul T. Schindler, President of AWF, "because poaching came to a halt almost overnight." The elephants have been given an opportunity to recover from the devastating massacres of the 1970s and 1980s. Experts believe it will take years for elephant populations to stabilize sufficiently to remove the elephant from the Endangered Species list, so we must keep a close eye on its status. We do that by continuing to foster appreciation for them and other animals everywhere.

II

Time check: We are roughly five years into the era of marketplace-as-solution-to-all-our-social-ills. Sometime around the turn of the decade, everyone agreed to equate democracy with having many choices at the checkout counter, and the "market solution" became a form of magic incantation, capable of transmuting pollution into gold and wastelands into wetlands.

Nowhere does this phenomenon arise more starkly than with the international trade in wildlife, particularly elephants and rhinos in sub-Saharan Africa. Trade in rhino horn has been banned by the Convention on International Trade in Endangered Species since the late '70s. [Trade in ivory has been banned since 1990.] The push is on, led by four African nations and supported here by some environmental groups, to reinstitute a limited trade. The reform rests ostensibly on the twin theories that, without the support of local people and without more money for government management programs (derived from the profits from the sale of animal parts), Africa's wildlife population will continue to decline.

A Dubious Proposition

That selling off Africa's wildlife will save it is a dubious proposition, but not atypical of the detached analysis that plagues this new school of thought. The notion founders on a pair of powerful illusions. The first is that all Africans, from rural farmers to members of parliament, from Tanzanians to Zimbabweans, share the same philosophies of life. The second is that this little exercise in applied economics can take place in isolation from the economic storms that are buffeting the region.

Animals are fast disappearing in parts of Africa that are economically devastated and, by consequence, profoundly corrupt. In Zimbabwe, according to research by the Humane Society of the United States (HSUS), bureaucrats in search of hard currency are stockpiling rhino horn and ivory by overestimating elephant populations and arguing for "culls," and by promoting ineffective rhino dehorning operations paid for by Western philanthropies. A government document obtained by HSUS investigators lists prices for Zimbabwe's wildlife, from $350 for a leopard to $200 for a male sable and $12 for a bat-eared fox.

© Whittemore/Rothco. Reprinted with permission.

The government of Zimbabwe argues that lifting the trade ban will provide income to manage the parks, but of the roughly $83 million that the country earned from park tourism in 1993, less than two percent went to protecting the animals. Last time Zimbabwe conducted a sale of ivory, it netted some $800,000, a pittance compared to what it obtained from non-lethal tourism, and a tiny fraction of the money that would be needed to police the parks effectively.

This is by no means an exercise in Africa-bashing. On the contrary, Africa has in general made a commitment to preserving its natural heritage that puts America's to shame. Aside from the state of Alaska, no region on Earth has set aside a larger portion of its terrain for wildlife protection. And few populations face more fundamental competition with wild creatures than rural Africans.

The failure here is in the attempt on the part of well-meaning environmentalists in the West to force the complex and desperate situation of rural Africa into the straitjacket of a "market" theory that bears little relation to reality.

Economic Support

If the United States wishes to help Africa save its wildlife, it will have to pay considerably more in broad-spectrum, enlightened economic support to the region. Sub-Saharan Africa is paying billions more in debt service than it receives in foreign aid right now, an astounding financial hemorrhage that, more than any other single factor, contributes to the pressure on elephants and rhinos. The support that does come in, in the form of loans for large-scale "development" projects, usually does more harm than good. And the United States will have to continue to oppose opening up markets in animal parts, while clamping down on the world's leading buyers, Taiwan and China, through the economic sanctions permitted by U.S. wildlife laws—sanctions which have never been used.

Ultimately, Africa's wildlife must be extricated completely from market considerations if it is to survive. This fact does not lend itself to the glamour of tough-minded "cutting edge" economic theories, but it offers some hope to the animals themselves.

> "There is no evidence that long term a ban in ivory will work."

The International Ivory Ban Should Be Lifted

Richard C. Morais

In response to the drastic reduction of African elephant populations throughout the 1970s and 1980s, the Convention on International Trade in Endangered Species (CITES) instituted a ban on ivory products that took effect in 1990. In the following viewpoint, Richard C. Morais contends that the ivory ban and other restrictions on the African wildlife trade are harmful both to impoverished Africans and to the wild animals. Morais advocates lifting the ban and instituting strict identification measures to ensure that ivory has been obtained legally. Allowing African communities to make money from their wild animals through safaris and trade, Morais argues, will benefit Africa's endangered species in the long run. Morais is a contributing editor for the financial magazine *Forbes*.

As you read, consider the following questions:

1. In the author's opinion, how do wild animals contribute to human starvation in Africa?
2. How can privatization of wildlife benefit both humans and wildlife, according to Morais?
3. According to the author, why did ivory prices drop after the ivory ban was implemented?

Richard C. Morais, "Save the Elephants!" *Forbes*, September 14, 1992. Reprinted by permission of *Forbes* magazine, © Forbes, Inc., 1992.

After you read this story you will understand why ordinary folks in Africa don't feel quite so romantic about wildlife as American city dwellers do.

One day the roar of rogue elephants woke Chief Sinagatenke from a sound sleep. Scrambling from their huts, he and his fellow villagers yelled and banged tin pails, but it was no use. The elephants tore the roofs off huts and grain storage units, eating a season's supply of food.

The huts could be repaired, but the food was a more serious matter for the villagers. Sinagatenke and the 664 families in his village in the Tyunga ward near Lake Kariba in northwestern Zimbabwe scratch a living out of the arid, rocky earth; they plant corn, sorghum, millet and watermelon. If there isn't enough food, people go hungry.

It often comes down, then, to a struggle between man and beast for food. If the elephants get at the grain, the children don't eat. "Wildlife is mostly responsible for our starvation," complains a Sinagatenke villager in dusty rags. Not just elephants, but all the creatures that swarm in the dense brush are wildlife: impala, baboons, wild pigs, eagles.

Destruction

During just a few months of 1992, Tyunga villages of a few thousand residents fought off dozens of attacks from hungry animals: buffalo, elephants, hippopotamuses. In March 1992 two grain storage units were pulled down by elephants in the neighboring village of Chief Siamupa; then a lion killed 25 goats in one night; a hippo returned six times to one field of corn. These attacks are calamities to a people suffering from malnutrition and protein deficiency, 85% of whose children have bilharzia, a painful urinary parasite.

"We live under house arrest," says a Tyunga elder standing in front of a circle of huts. . . . "People can only move safely between 7 a.m. and 6 p.m. At night we shut ourselves in our houses."

Animal conservationists rarely address the price people pay living with wild animals. They are appalled when African villagers slaughter wild animals. Perhaps they would be more understanding if they shared some of the villagers' hunger.

But out of Africa comes new hope that ways can be found to reconcile the conflicting needs of man and wild animals.

In 1990, in 12 painfully poor rural districts of Zimbabwe, the government turned over to local villages title to the animals that roamed over communal lands. Suddenly, the villagers found that what they had thought of as a liability could be turned into a valuable asset.

"The idea is that the communities that pay the price of living with wildlife [should] also reap its financial rewards," says Keith

Madders. He's the London-based director of Zimbabwe Trust, a lean, no-nonsense charity that is helping to implement the program in the villages. "Unless you give wildlife economic value, convincing locals [that providing for] it is the best use of their land, wild animals will not survive."

Selling Safaris

Put simply, the villagers have found how to earn money, not from hides and tusks, but from outlanders hungry for the sight of real wild animals. In 1991 Tyunga's local council put up for bid the region's hunting rights. That year a safari operator paid $63,600 for the franchise. In 1992 there was a bidding war and the hunting package went for $350,000. You can buy a lot of food for that much money in Africa.

Reprinted with permission of Chuck Asay and Creators Syndicate.

European and American hunters have been forking over as much as $40,000 for the privilege of shooting, say, an elephant or some other species, such as a wildebeest, an impala or a warthog. In the past such money disappeared into Zimbabwe's treasury. From there? In Africa, often as not, into private pockets. The villagers rarely saw a dime. "Now everybody realizes the potential of wild animals," says the chairman of Communal

Areas Management Program for Indigenous Resources—or Campfire, for short—as the privatization effort is called.

In June 1992 Chief Sinagatenke's village got $1,000, its share of the first Campfire cash installment for that year. In a raucous democratic process, his village decided to invest $747.60 in grain, to get the village through the drought; to put $98 into a bank account; to use $22 to transport village-elected Campfire officials to district-level meetings; the rest went on a village celebration. This reporter watched the festivities sitting under a torn tarp with elders and chewed on salty pieces of buffalo and goat. "We no longer wait for handouts," said the local Campfire chairman, who is known simply as Mudenda. Verbally and in writing, all the funds were accounted for to the penny.

It's the old story: People will almost always take better care of their own property than they will of someone else's—a lesson the former socialist countries have learned the hard way. Whereas killing animals was an act of survival for the villagers in the past, today it amounts to destruction of their own property. So they find ways to protect their homes without killing animals. Recently, families in Chief Sinagatenke's village secured themselves behind an electric fence funded by the U.S. Agency for International Development. The 14-mile-long fence protects some 19 square miles of land with 6,500-volt batteries charged by solar energy. "Elephants are learning that the fence shocks and are beginning to fear it," wrote G. Moyo, a fence minder, in a report.

The village is spending its own money to maintain the fence. The Tonga tribespeople have voluntarily put themselves and their crops behind it, allowing the wildlife to roam free. What a change in worldview for these people.

The Success of Privatization

The principles behind Campfire are not new. They were first introduced back in 1975 when Zimbabwe was still white-ruled Rhodesia. In a move unusual for Africa, Ian Smith's government granted large landholders ownership rights to the wildlife on their property. A robust game hunting market quickly developed. When President Robert Mugabe's black government took over in 1980, he wisely continued the program. But it was restricted to private land. The wildlife on the vast acreage of communal land was still considered the property of the central government, and it was Harare [the capital of Zimbabwe] that reaped the cash harvest.

But even with this limited application of the privatization principle, Zimbabwe's elephant population is up to 77,000 now, from 32,000 in 1960. In the rest of Africa, where wild game belongs to the state, the story is different. Poachers halved the elephant population in Africa from an estimated 1.2 million in 1981 to 623,000 just eight years later.

And yet such is the statist orientation of most international environmental organizations that they continue to promote failed policies. In 1989 animal conservationists backed by sympathetic but ignorant media coverage in the West convinced the Convention on International Trade in Endangered Species in Flora & Fauna (CITES), an international body, to ban all trade in elephant products. At one stroke of the pen consciences were appeased. The West had done its part in saving the elephant.

But the depredations on poor villagers continued. And so did the temptation for them to hunt illegally for sale to smugglers.

"There is no evidence that long term a ban in ivory will work," says André DeGeorges, an environmental consultant in Africa. According to *Elephants, Ivory and Economics*, an in-depth study by four respected British academics, the crash in the price of ivory was the result of massive stocks' being liquidated just prior to the ban, and a genuine drop in demand for ivory in Europe and the United States because of the adverse publicity.

The bulk of the ivory market is not there, however, but in Asia, where ivory stamps are a status symbol. In the East there is little sign the ban has reduced demand, say the academics.

In Ndebele, a principal language of Zimbabwe, the word for wildlife (*nyamazana*) means "the meat that walks." It is naive to think that Africans on the edge of starvation will not satisfy Asian demand for ivory and will not use weapons to defend their children's food against marauding animals.

In short, the politically correct solution—government action—doesn't work. In 1970 there were 65,000 black rhinos in Africa; today there are probably fewer than 4,000. A total ban on black rhino horn has been in effect for almost two decades.

These bans on ivory and other such products, so beloved of environmentalists who do not have all the facts, make little sense. African national parks routinely gather tons of ivory every year from elephants dying of natural causes or from culling. Ivory could be carefully tagged and identified as coming from animals that died natural deaths so that poachers' ivory couldn't be laundered. The legal trade of ivory would raise sorely needed cash for the national parks trying to save wildlife.

Brian Child, chief ecologist at Zimbabwe's Department of Parks & National Wildlife Management, who sometimes hitchhikes into the bush because of budget constraints, estimates that legal trade would increase his budget tenfold overnight.

The free market is a wonderful instrument for getting things done without coercion and without corruption. Africa cannot afford an environmental colonialism that despises market mechanisms and private property.

5 VIEWPOINT

"Ecologists hope to convince those who live in and near rainforests that the trees and plants . . . can provide a continuing source of income."

Buying Rain Forest Products Preserves Biodiversity

Diane Jukofsky

The tropical rain forest is an important source of biodiversity; scientists believe it is home to at least half of all living species on Earth. In the following viewpoint, Diane Jukofsky examines the attempts of U.S. and British companies to save this ecosystem by means of a marketing trend called sustainable harvesting. According to these companies, Jukofsky relates, logging of rain forests can be prevented if indigenous peoples and local farmers can make a steady income by harvesting nuts, oils, and other renewable forest resources. By purchasing products that contain ingredients that were responsibly harvested from the rain forests, Americans can help preserve this important ecosystem, she concludes. Jukofsky is the codirector of the Rainforest Alliance's Latin America office in San Jose, Costa Rica.

As you read, consider the following questions:

1. According to Jukofsky, what types of products have their origins in tropical rain forests?
2. What is product sustainability, as defined by Jukofsky?
3. According to Jason Clay, quoted by Jukofsky, in what way does farming cupuaça save rain forests?

From Diane Jukofsky, "Can Marketing Save the Rainforest?" *E: The Environmental Magazine*, July/August 1993. Reprinted with permission from the Rainforest Alliance and *E: The Environmental Magazine*; Subscription Dept.: PO Box 699, Mt. Morris, IL 61054. (815) 734-1242. Subscriptions are $20/year.

Kathryn Alexander saw the green light when she read an article about Cultural Survival, the indigenous-rights group based in Boston. "I was truly inspired," she remembers. Alexander is president of Tropical Botanicals, a California company that produces such sexy-sounding products as "Passionflower Massage Oil" and "Babaçu Nut Body Lotion."

What inspired Alexander to start Tropical Botanicals and its line of rainforest bath products was Cultural Survival's program of buying directly from harvesters in tropical countries, then selling these fruits, nuts and oils to manufacturers. Like other companies that buy from Cultural Survival, Tropical Botanicals pays a five percent premium above market cost, in their case for oil from the babaçu palm tree that it whips into a fragrant, creamy lotion. Cultural Survival then turns profits back to the South Americans who gathered the palm fruit.

Use It or Lose It

Alexander is not alone in the rainforest products market. Dozens of beauty products and munchies promise consumers that their purchase is "actually helping to preserve the precious rainforests," as it says in the Tropical Botanicals ad, just to the right of the handsome jaguar.

The concept, roughly, is "use it or lose it," a far cry from when many environmentalists thought that the best way to save wild habitat was to declare it off-limits to the destructive habits of humans. Now, ecologists hope to convince those who live in and near rainforests that the trees and plants in their backyards can provide a continuing source of income.

In 1989, Charles Peters, of the New York Botanical Garden's Institute of Economic Botany, completed a study that finally provided rainforest conservationists with the data they needed to prove that a rainforest is worth more intact than logged or burned for cash crops or cattle pasture, the usual fate of the 50 million acres of tropical forest lost worldwide each year.

Peters and two colleagues found that the fruits and rubber that could be gathered from one hectare (2.5 acres) of Peruvian rainforest could net a harvester $422 a year, *every* year, which adds up to a far more secure future than the $1,000 that hectare would yield, just once, if it were cut for timber.

Now, in Guatemala, Peru, the Philippines, Ecuador, Zambia, Indonesia—in nearly every country where there are still large stands of rainforest remaining—experts are combing the forests to see what might be harvested and sold.

The number of products with origins in the rainforest already on the market seems remarkable—until you consider that half of all the species on Earth are found in tropical forests. Oranges, lemons, bananas, pineapples, chocolate, coffee, avocadoes, resins used in paints and varnishes, latex, bamboo, rattan, many dusky-hued hardwoods used to make furniture, and more than 40 prescription drugs have their origins in wild tropical plants.

Valuable Plants

People who live in and near tropical forests regularly use hundreds of plant species Westerners have never tasted, smelled or seen. The 200 families who live along the Río Capim in Brazil's Amazonia, for example, are "utterly dependent" on plants, says Woods Hole Research Center ethnobotanist Patricia Shanley. "Their daily routines demand an intimate knowledge of the surrounding forest." Not long ago, they asked Shanley and her colleagues for help in determining which plants in their forests might be marketable. Shanley surveyed the nearby forests and found 400 different plant species. She found trees like bacuri, ushi and piquia, whose fruits are in demand at local markets, and andiroba and copaiba trees, which yield effective medicinal oils. While these products may have value internationally, Shanley is concerning herself right now with what their worth might be in local and regional markets. "The closest market is three hours away by boat," she explains, "and selling a plant is a whole new concept to people here."

It's a concept that the Río Capim villagers and hundreds of others who live near the forest's edge are learning: There are people willing to pay for the bounty in their backyards. Buyers may be downriver, or on another continent.

The members of the Xapuri Agroextractive Cooperative are rubber tappers in the state of Acre in Brazil. To supplement the income they earn from the latex tapped from rubber trees in the Amazon, they now gather the fruit of Brazil nut trees from the forest floor. Inside each softball-sized fruit are up to 20 nuts, which are brought to a nearby processing plant. There the nuts are dried, shelled and sorted for shipment to foreign markets.

The Birth of a Movement

In 1989, Cultural Survival bought a shipment of Brazil nuts from the rubber tappers and sold them at five percent above the market price to a new company called Community Products, which used them in a nut brittle it dubbed "Rainforest Crunch." Cultural Survival returned 100 percent of the profit from the sale to the Xapuri cooperative. Suddenly, a movement was born. According to recently departed director Jason Clay, who founded the organization's rainforest products marketing program, Cultural Survival now sells 15 commodities from 10 to 14 different countries to 66 companies. The program has thus far generated about $3 million. All profits are returned to the product harvesters. (Clay is now co-director of the Washington, DC-based group Rights and Resources.)

New England Natural Bakers is one of the companies that buys from Cultural Survival. President John Broucek started the company in 1989, "after I read an article about Jason Clay and Community Products."

Encouraging Reforestation

Cupuassu trees have been planted as part of a reforestation program in the western Amazon, where thousands of colonists have deforested the area in mostly unsuccessful attempts to develop cattle ranches and farms. By developing and expanding the market for cupuassu, Cultural Survival Enterprises is helping to make the sustainable production and harvest of this fruit profitable and therefore encourage the colonists to remain on the reforested land rather than clearing more land for new farms or ranches.

Peggy Eulenson, *Environmental Action*, Fall 1993.

Broucek first introduced "Save the Forest" granola cereals, then fruit and nut snacks. "We source 95 percent of our products from Cultural Survival," says Broucek. "Any of our ingredients that come from rainforest areas, we buy from them," even though he could be saving a good deal if he bought his Brazil nuts from another source.

Indeed, there's a glut of Brazil nuts on the open market now. "The current market price is about 85 cents a pound," says Clay, "and we're paying about $1.25. . . . Many of the 'green' companies just started. . . . It hurts them to pay so much more than the market price.". . .

Rachel Perry Body Care Products has sold rainforest products since the 1970s. In fact, the company has even trademarked the word "rainforest." Three years ago, as Cultural Survival and Rainforest Crunch were making news, Rachel Perry introduced Rainforest Botanical Therapy Body Care products. "We thought that by creating this product line it would help call attention to the rainforest situation," says marketing manager Melinda Rubin.

Sustainable Harvesting

Rachel Perry buys Brazil nut oil, babaçu, copaiba, agave, banana nut, coconut and orchids from Cultural Survival and other sources. "We ask that our sources provide us with substantiation that these ingredients were sustainably harvested," explains Rubin.

Like other Cultural Survival customers, Rachel Perry and New England Natural Bakers try to be careful about the claims they make. "We say our goal is to use as many sustainably harvested products as possible. We don't say every product is sustainable," says Rubin.

Product sustainability—or harvesting without damaging the ability of the plant that produced it to produce again—is one of the controversial aspects of rainforest marketing. Clay maintains that the only way to develop sustainable sources is to buy first from the open market—that is, from commercial sources—even though it means that, initially, he really doesn't know how the products were harvested, or by whom or under what conditions. "We use that as a way to get companies interested and convinced that we can supply them with the quantities they want," he says. "All of our initial 15 commodities were commercially sourced. Now only 4 are part commercial, part local."

One product Cultural Survival still buys from commercial sources is annatto, a seed that produces a tawny oil that is used in Tropical Botanical's Annatto Natural Bronzing Oil. "We can't guarantee that what comes from commercial sources is sustainably harvested," Clay admits. "We *can* guarantee that the money generated from that sale goes back to set up" a sustainably harvested system.

Even products that are bought from local sources are not necessarily gathered from the rainforest. Cashews, for example, a common ingredient in many rainforest cereals and snacks, come from trees grown in plantations. They are farmed like any crop, planted in rows on land where rainforests once flourished.

"A lot of the products being sold as rainforest products are actually farmed," says Michael Balick, who directs the Institute of Economic Botany. "Some of the products in the stores are from plants that are not sustainably extractable, yet they are being sold as contributions to the movement."

Orchards or Pastures

Clay makes no apologies. "We buy cupuaçu from a cooperative near Belem in Brazil," he says. The cooperative's members, says Clay, are recent colonists who moved in and cleared rainforest. Cultural Survival funds helped set up a small processing plant that will buy the fruit from farmers in the area. "If we can't figure out a way for the colonists to make a decent living, they're going to keep cutting down the rainforest and sell it as pasture," Clay warns.

Knudsen & Sons, Inc., which buys cupuaçu from Cultural Survival and sells it as one of the company's Tropical Rainforest Juices, describes the fruit as tasting something like a honeydew melon. It grows wild, but many farmers like those in Belem are planting cupuaçu to take advantage of the fruit's new international market. "There are so many people planting cupuaçu," says Woods Hole botanist Shanley, "that it could be a time bomb. The fruit could flood the markets, and that could be the worst thing in the world, because finally these farmers are taking a risk and planting something new."

Michael Saxenian of the environmental group Conservation International (CI), which also works in the production and marketing of rainforest products, shares Shanley's concerns. "We believe that simply taking non-timber products out of the market and selling them is not enough to guarantee that our conservation objectives will be met," he explains. "We work with scientists to study the ecological sustainability of any product that we offer so, while it's always difficult to determine what is a sustainable product, we think we have as sound a scientific basis as possible in making sure that we're not exceeding those limits." CI first identifies what it calls a "hot-spot" ecological system, a natural area rich in biodiversity that faces development threats. Working with the people nearby, CI designs programs that involve community development, conservation science, long-term management and policy work.

CI's first forest product extracted under this multi-faceted program was tagua nut, the hard seed of a rainforest palm tree. Tagua nut, sometimes called "vegetable ivory" for its creamy white color and easily carved texture, is sold as buttons, jewelry and carvings. CI linked up with a community development group outside the Cotacachi-Cayapas Ecological Reserve in Ecuador to establish a locally run business that purchases tagua and re-sells it to factories on the coast, which in turn shape the tagua into disks for buttons.

"First," says Saxenian, "we did an ecological analysis of the area to determine whether tagua nuts could be sustainably harvested. We also provided capital and some of the infrastructure the community enterprise needed, like scales for weighing the tagua nuts, and boats for transporting them."

CI then found three U.S. button manufacturers who would buy the semi-processed tagua disks, as well as companies who would buy the finished buttons. In the first year of CI's "Tagua Initiative," seven million buttons worth $500,000 were sold. CI is developing similar programs with a variety of products in ecological hot-spots in Colombia, Guatemala, Peru and the Philippines. . . .

Saving the Rainforest

Satisfied that the ingredients were harvested in a responsible way, and that the profits will benefit, if not an indigenous tribe, at least deserving farmers or rubber tappers and their families, consumers may still wonder, "Just how many acres of rainforest is my purchase going to save?"

"It's difficult to say," acknowledges Saxenian. "We do have anecdotal evidence that suggests we are having a real impact. A 12,000-hectare [30,000-acre] banana plantation was proposed for the site where we are working. The Tagua Initiative was an important topic in the debate in the community . . . and the plans for the plantation were scrapped."

Notes Jason Clay: "I can't say with any kind of conviction that we've saved a hectare of rainforest [but] we have generated a lot of income for as many as 20,000 people."

Many rainforest product companies donate a portion of their profits to nonprofit groups that support rainforest conservation, chalking up yet another helpful gesture in their favor. Consumers may want to read labels to find out how much is donated, as well as the names of the groups that receive funds. If you're unfamiliar with a group's goals, request more information.

And then, make that purchase. Although, as John Broucek of New England Natural Bakers notes, "The efforts of all these companies put together are not going to have a dramatic effect," the majority of rainforest product companies and organizations are led by people with sincerely good intentions, who are trying to merge heart and business in a positive way. It's a step toward maintaining the forests for the future.

6 VIEWPOINT

"International marketing has usually resulted in products' being over-exploited to the point of extinction."

Buying Rain Forest Products Does Not Preserve Biodiversity

Stephen Corry

Since 1989, several U.S. and European companies have marketed products that are advertised as containing ingredients harvested from the rain forest in an ecologically responsible manner. In the following viewpoint, Stephen Corry argues that many of the claims of these companies are deceptive. Purchasing their products neither benefits indigenous peoples nor contributes to the preservation of the rain forests by encouraging alternatives to logging, Corry maintains. Furthermore, he warns, some of the companies plan to eventually include rain forest timber among their products, which would endanger the very forest they purport to be saving. Corry is the director general of Survival International in England.

As you read, consider the following questions:
1. How does the marketing of rain forest products resemble colonial domination, in Corry's opinion?
2. According to the author, what types of false claims about rain forest products have been made by businesses?
3. What residual effect will the rain forest product trend have on marketing, in Corry's view?

From Stephen Corry, "The Rainforest Harvest," *The Ecologist*, July/August 1993. Reprinted with permission.

Of all the planet-friendly advertising slogans that have emerged over the last few years, none has been more evocative than "rainforest harvest". The concept links the exotic luxuriance and the impeccable environmental credentials of tropical forests with ancient and reassuring associations of fertility and abundance.

The theory behind the rainforest harvest is simple: if it can be shown that forests are of more value when left standing than when they are felled, then they are more likely to be preserved. This value is expressed in the monetary price of products which can be extracted from forests, mainly fruits, nuts and cosmetic oils—the subsistence value of traditional forest products is of no account. Proponents of the harvest claim that it will assure forest dwellers an income as producers of raw materials for North American and European consumers, and that they will become more secure and empowered as a result.

The idea has been given widespread publicity since 1989 when it was first sold to the press as a key to the future for both rainforests and tribal peoples. Two companies, in particular, have been involved, Cultural Survival Enterprises in the United States and The Body Shop in the United Kingdom. Both companies have extolled the virtues of the rainforest harvest in glowing terms and compared it favourably with other projects. According to J. Forster, "Trade is much better than a handout and it will be far more effective at protecting forest people. . . . It's good business, not just for business, but for human rights and the environment." Or more recently: "One of the best ways to help indigenous groups preserve their native lands is to discover markets for . . . products," according to J. Christensen.

But strip away the advertisers' hype, subject the idea to serious scrutiny and a different picture emerges. So far, the harvest has done nothing to empower indigenous people or to protect forests; instead it has diverted the public's attention from more critical matters, including the demand for indigenous land rights.

The Colonial Market

Marketing forest products is by no means a new idea—throughout the colonial era tropical forests were regarded as a source of raw materials for the European market. But far from encouraging conservation, international marketing has usually resulted in products' being over-exploited to the point of extinction. For example, the extraction of rubber and ivory—the basis for the brutal colonial domination of the Congo basin in the late 19th century—all but eradicated the rubber vines and elephants over vast areas. In South-East Asia, where, particularly since the early 1970s, the trade in hornbill ivory, rhino horns, bear paws, bezoar stones, gaharu incense and birds' nests has led to over-extraction and eventually the local extinction of species, and so the demise of the trade itself.

The history of colonialism shows that rural inhabitants, including indigenous peoples, have not become more secure or empowered by supplying raw materials for foreign markets. At best they have become exploited and dispossessed of their lands; often they have simply been exterminated. . . .

Unfounded Claims

The principal harvest product—the one which has received most of the press attention—is Brazil nuts. The flagship of the harvest, indeed the only product which has become at all well known, is "Rainforest Crunch", a candy bar containing Brazil nuts from the Brazilian tropical forest and many other ingredients which have nothing to do with rainforests.

Rainforest Crunch was originally sold with the following claim on its packet: "The nuts used in Rainforest Crunch are purchased directly, with the aid of Cultural Survival, from forest peoples." In fact, this was not true. The Brazil nuts used do come from rainforest areas, but for two years or so, all of them were bought on the normal commercial market.

A Threat to Indigenous Peoples

The harvest projects, if successful, could encourage more colonists to seek a living in the rainforest, further marginalizing tribal peoples. Colonization of the rainforests, combined with forced relocation due to large scale development projects (including export agriculture), have been the major source of destruction of tribal peoples' lands and communities, exceeding, many believe, the negative impacts of logging and mining. Colonizers force indigenous peoples off their lands, driving them deeper into the forest, and bring violence and diseases to which the Indians have no resistance. Non-indigenous people are desperate for ways of digging themselves out of poverty and will flock to the forest to harvest and process (however minimally) these goods.

Susan Meeker-Lowry, Z *Magazine*, July/August 1993.

The Brazil nut industry is a well-established extractive business (with a turnover of $20 million in 1989) which relies on an unskilled and poorly paid labour force, and is dominated by the wealthy and powerful, not by indigenous communities. Workers' rights, minimum wages and unionization are all ignored or suppressed. Some of the nuts for Rainforest Crunch were acquired from exploitative suppliers who saw their profits increase as a result.

Cultural Survival's justification for making ethical claims for a product derived from commercial and ethically dubious sources was that it first had to create the market for rainforest products; and that to do so it needed large quantities of the nuts, more than any "ethical" supplier could provide. The market has to be created because the product is entirely dependent on its publicity, on the invention of a new "need". Consumers are not really interested in buying a Brazil nut candy bar for itself, they buy it because helping to save forests makes them feel good. But in this case the "feel good" factor is a fake.

Even now, four years after the scheme started [1993], none of the product comes from indigenous peoples. Some of the Brazil nuts are collected by non-indigenous rubber-tappers in the state of Acre. Yet Cultural Survival describes itself only as working on indigenous, tribal peoples' or ethnic minorities' issues—in none of its promotional literature does it include in its objectives any mention of helping non-indigenous communities, let alone suppliers and brokers. In fact, even non-indigenous forest people do not often benefit directly. Brazil nuts and copaiba oil "are the only important supplies coming directly from forest people. Commercial suppliers and brokers provide most of the rest of what Cultural Survival imports," according to P.J. Jahnige.

The term "forest peoples" is still used in recent packaging, but the claim has been watered down. It now says: "Profits from the nuts we buy are being used to develop small Brazil nut processing factories that are cooperatively owned and operated by forest peoples." There is no mention of who supplies the nuts used. The packaging for one of the several ancillary products now available, "Rainforest Crunch Popcorn", makes even vaguer claims, stating simply: "Thanks for helping us save the rain forest by buying this butter crunch nut popcorn. The Brazil nuts grow wild in the rain forest."

Deluding the Public

Harvest advocates argue that the labels on their goods are an important educational tool. They say that they "use product packages to educate consumers about both rain forests and the peoples who live in them. In 1991, some 30 million Americans bought products that explained the importance of the rain forests, how consumers could help local groups protect their resources," according to Cultural Survival. Unfortunately the cocktail of half-truths and advertising clichés that appear on their products suggest that the public may be being deluded rather than enlightened. . . .

The enthusiasm for "harvest" products may well prove to be a short-lived fad and eventually fizzle out. Even if this is the case, it will doubtless leave a residual legacy; rainforest hype, like other forms of green, will continue to be used in advertising and packaging.

In this respect, an aspect which may be peripheral now could prove central in years to come. Advocates of the harvest play heavily on another piece of jargon, "Non-Timber Forest Products", which in practice means mainly fruits and nuts. But they have occasionally let slip that they believe timber could eventually play an important role in their schemes. They tend to keep this to themselves because many of the organizations they are trying to seduce are fervently opposed to logging. The barely whispered message is more or less identical to that peddled by timber importers, who have started their own campaign, called "Forests Forever", to counter the fierce criticism which has been mounted against their activities over the last few years. The timber traders say that judicious felling can preserve forests and is therefore environmentally sound.

Logging the Forest

It may be that careful logging could extract timber of greater monetary value than any fruit or nuts, without entirely destroying the forest. But there are enormous problems: no one has any idea if tropical rainforests can be logged sustainably or not; and logging is not and will never be carried out "judiciously and carefully" in countries where controls are ignored and corruption starts at the top. For example, British mahogany importers are hiding behind Brazilian certificates attesting that their wood is not being taken from conservation zones or Indian reserves. These are falsified certificates: most of the imported timber now comes from Indian areas.

Harvest advocates say that higher prices paid for forest produce will promote conservation. But higher prices for timber translate into more intensive cutting and make extraction from more remote areas financially attractive. They also result in the construction of more roads, and where the timber roads go, colonization and devastation soon follow. As Cultural Survival itself admits, "the development of markets for sustainably harvested commodities and the destruction of rainforests . . . both depend on . . . improved transport".

Timber production is the inevitable conclusion for those who pursue the logic of the rainforest harvest.

Periodical Bibliography

The following articles have been selected to supplement the diverse views presented in this chapter.

Sharon Begley	"Killed by Kindness," *Newsweek*, April 12, 1993.
Raymond Bonner	"Crying Wolf over Elephants," *The New York Times Magazine*, February 7, 1992.
Peggy Eulenson	"Whose Rainforest Is It Anyway?" *Environmental Action*, Fall 1993.
Sidney Holt and Johan Joergen Holst	"What Is Sustainable Development? The Test Case over Whaling," *Our Planet*, vol. 5, no. 5, 1993. Available from the United Nations Environment Programme, PO Box 30552, Nairobi, Kenya.
Michelle E. Howard and Clifford E. Thies	"The Invisible Hand That Has, Time and Again, Saved the Whales," *St. Croix Review*, February 1995. Available from PO Box 244, Stillwater, MN 55082.
Daniel R. Katz	"Alternatives Needed to Save Rain Forests," *Forum for Applied Research and Public Policy*, Winter 1992. Available from 1004 Mississippi Ave., Davenport, IA 52803.
Joe Kirwin	"Policing the Wildlife Trade," *Our Planet*, vol. 6, no. 4, 1994.
Brad Knickerbocker	"Illegal Trafficking in Wildlife Persists," *The Christian Science Monitor*, March 8, 1993. Available from One Norway St., Boston, MA 02115.
Eugene Linden	"Sharpening the Harpoons," *Time*, May 24, 1993.
Susan Mekker-Lowry	"Killing Them Softly: The 'Rainforest Harvest,'" *Z Magazine*, July/August 1993.
Cynthia Moss	"Elephants Slaughtered," *HSUS News*, Spring 1995. Available from the Humane Society, 2100 L St. NW, Washington, DC 20037.
Daniel Stiles	"Harvesting the Forest," *Our Planet*, vol. 6, no. 4, 1994.
Langdon Winner	"Kill the Whales?" *Technology Review*, November/December 1994.

CHAPTER 5

Are Humans an Endangered Species?

Endangered Species

Chapter Preface

The tropical rain forests of Sarawak, a Malaysian state on the island of Borneo, are home to the Penan, the Iban, the Kenyah, and many other indigenous, forest-dwelling peoples. However, as Robert Weissman writes in the *Multinational Monitor*, "in recent years, nearly 3 percent of Sarawak's primary forests have [been logged] annually." Logging operations have also resulted "in the contamination of 70 percent of Sarawak's rivers and streams," according to Art Davidson, author of *Endangered Peoples*.

As water pollution and soil erosion affect fish and animal populations, Davidson argues, the native peoples of Sarawak "lose their traditional sources of food and can no longer live by roving through the forest. Many of them are now crowded into squalid camps, where they are succumbing to parasitic infections, dysentery, [and] tuberculosis." In the forests of the Amazon, also affected by logging, the indigenous population has declined from four million in 1900 to approximately 750,000 in 1990. As indigenous populations decrease due to environmental destruction and loss of native lands to development, many anthropologists warn that unique native cultures are becoming extinct.

Others feel that, by leaving behind some of their traditions for modern civilization, indigenous peoples can gain a more prosperous life. Tan Sri Taib, the chief minister of Sarawak, told the *Asian Wall Street Journal* that the logging of the state's forests had "benefited a lot of people. You can see Ibans and Kenyahs becoming truck drivers; some are earning as much as people with B.A. degrees." Officials in many developing countries maintain that native peoples should adapt to modern life not only on their own behalf but to help advance the nation's economy and technology. In May 1992, Rafidah Aziz, Malaysia's minister of international trade and industry, said of the Penan tribe:

> We'd like to take them out of the jungle. Give them a decent modern living. It has nothing to do with logging, actually. . . . We're talking about the twenty-first century. We cannot afford to have some of our population still hunting monkeys. . . . *We're* talking about children using computers.

Proponents of modernization suggest that indigenous cultures are not becoming extinct but rather are naturally evolving from a primitive way of life to an advanced one.

Humankind's ongoing transition from living closely with nature to industrialized urbanization has led many experts to debate how best to ensure the survival not only of indigenous peoples but of all humanity. Whether humans themselves have become an endangered species is debated by the authors in the following chapter.

VIEWPOINT 1

"Most of the Amazon's Indian societies are already extinct, and those that remain face imminent destruction."

Indigenous Peoples Are Endangered

Art Davidson

Art Davidson is the author of numerous books, including *The Circle of Life, Does One Way of Life Have to Die So Another Can Live?*, and *Endangered Peoples*, from which the following viewpoint is taken. Relating the problems faced by the native peoples of the Amazon, Davidson warns that indigenous cultures are highly endangered. Disease, pollution, and the destruction of native habitats are eradicating entire cultures and decimating tribal populations, he contends. Native cultures have much to teach the modern nations, Davidson maintains, but this wisdom is rapidly being lost as these cultures become extinct.

As you read, consider the following questions:

1. According to Davidson, what is the rate of cultural extinction in Brazil?
2. In the opinion of Evaristo Nugkuag, quoted by the author, how does the deforestation of the Amazon impoverish his people?
3. What are the causes of the two waves of death suffered by the Yanomami, in Davidson's view?

Excerpted from *Endangered Peoples* by Art Davidson. Copyright ©1993 by Art Davidson. Reprinted by permission of Sierra Club Books.

"Civilization must advance," read an 1854 report to the U.S. Congress on the Amazon's potential, "though it tread on the neck of the savage, or even trample him out of existence."

To explorers, missionaries, and all sorts of fortune seekers, the vast reaches of the Amazon have long beckoned as a last frontier waiting to be civilized. But for the people of the rain forest, the Amazon is simply their home. In the quiet light under the forest canopy, the rhythms of day-to-day life still play out much as they have for thousands of years. Over centuries of trial and error, these people have evolved ways of growing vegetables, hunting and fishing, and gathering wild fruits and nuts that allow the forest to replenish itself.

In recent years, images of the Amazon going up in smoke have jarred us into realizing that the rain forest is finite and fast disappearing. Fifteen percent of the Amazon is already deforested and an estimated twenty million acres of virgin forest are cleared annually. But what of the people?

Vanishing Cultures

Since the turn of the century, indigenous cultures in Brazil have been disappearing at the rate of one per year. As many times as I've read or reiterated this fact, I don't believe I've begun to fully comprehend the suffering it represents. It might be reassuring if we could relegate this enormous loss in the human family to travesties of the past. But we can't. If anything, the pace of cultural extinction in the Amazon may be quickening, as new roads are pushed through indigenous territories, bringing new waves of miners, settlers, alcohol, and disease.

"The indigenous societies in these tropical regions are becoming extinct at an even faster rate than the regions they have traditionally inhabited," says Jason Clay of Cultural Survival. "In many indigenous societies undergoing rapid change, young people no longer learn the methods by which their ancestors maintained fragile regions. Little time remains to salvage this knowledge."

Most of the Amazon's Indian societies are already extinct, and those that remain face imminent destruction. But is the extinction of these indigenous peoples inevitable? Brazilian law actually provides the assurance that native communities receive "permanent possession of the land they inhabit, recognizing their right to exclusive use of the natural wealth." However, another statute gives the government the right to expropriate Indian lands and relocate entire communities in the interests of national security or development. To provide the kind of security that allowed them to survive and evolve over millennia, native lands must be demarcated and then those boundaries rigorously guarded. Brazilian governments will, of course, come and go, some with less corruption and more commitment to indigenous peoples than others.

If they begin to accept, really accept that indigenous peoples are part of the environment, conservationists could play a pivotal role in the future of Amazon cultures. Each year, millions of dollars are spent to save endangered birds, mammals, insects, and snakes, but as yet there is only a minimal awareness that people can be endangered, too. . . .

Indigenous Struggle

It is difficult to meet Eliane Potiguara and not notice her commitment to the struggle of indigenous women. As president of Grumin, an organization of five hundred indigenous women, Eliane directs a number of initiatives. Pottery is made to enable communities to earn at least a small amount of money. One education program offers training for Indian teachers, another prepares classroom materials based on indigenous realities. She also runs a campaign to curb alcoholism. "All of our men drink," she says. "They feel impotent because they have no prospects for work and they see their communities dying."

Eliane's own Potiguara people (a person's last name is often the name of his or her tribe) still number about seven thousand. In the past twenty years, they have suffered one setback after another. They always had enough land for their needs until the government parceled half of it out to develop sugar-cane plantations. Freshwater shrimp was once a staple in their diet, but now miners have polluted the river and it is unsafe to eat the shrimp. "Whether they mean to or not, all the people coming to our region play a part in destroying the way we live," Eliane told me. "We have lost our land. There is no work for the men. Babies are dying from lack of food. People feel helpless. We are caught in the chains of hunger, misery, drinking, violence, and suicide.". . .

"There are many different ways to understand and to measure development," says Evaristo Nugkuag, president of COICA, the coordinating organization for indigenous groups of the Amazon Basin.

> We indigenous peoples have long had many different development models imposed on us and our territories by both the State and the private sector, and we have suffered enormously. If you want to know what development means to us, you must be willing to accept that our mode of development is not the same as yours. Our development is not based on accumulation of material goods, nor on the greatest rates of profit, obtained at the expense of our territories and future generations. . . . For us, development must take into account the future of an entire people.

The key to development from this indigenous perspective is having an extensive and diversified territory where all people, animals, trees, and rivers will share the benefits. "We were taught by missionaries and development workers to raise cattle," says Evaristo. "But we are beginning to realize that we made a poor deal: We sacrificed thousands of useful and potentially useful species just to feed a cow."

Long-Term Loss

"There is no need to overly romanticize indigenous cultivators," cautions Jason Clay. "Some have been known to forget where they have planted seeds and needlessly replant the same area." But as Clay emphasizes, peoples of the rain forest are adept at "reading the environment," and their land-use methods "are based on the view that the environment is the source of life for future generations and should therefore not be pillaged for short-term gain and long-term loss."

There can be no life if our forests are destroyed, says Evaristo. "We want to continue living in our homeland. We have no interest in taking everything the forest has to offer and moving to the city to live in material comfort from the profits of our plunder."

Every year, half a million acres of forest go up in smoke in the Peruvian Amazon alone. To the Peruvian government this spells progress. To the Indians it is, in Evaristo's words,

> a disaster that is impoverishing our people. To this day, hundreds of families are kept as slaves on farms or sent to cut timber, with no pay, no food, but with armed guards. Young girls are kept captive in the landlord's house to service him at his will. Both young and old are mutilated and beaten for trying to escape. Tuberculosis is rampant. The local authorities know this is going on, but they close their eyes.

> Slavery still exists—right here at the end of the twentieth century. We live in crushing poverty. The authorities have a total disdain for our rights. Our forest has become a land of violence. Poverty, corruption, injustice, and lack of basic human rights all generate violence. And that is why the Amazon is so violent today.

Waves of Death

Prior to the 1970s, the Yanomami dwelled beyond the reach of the twentieth century in a remote area of northern Brazil and southern Venezuela. There were about twenty thousand Yanomami then, all living in small community groups that moved every four or five years to give the forest and their garden soils time to regenerate.

In 1973, the world of the Yanomami changed abruptly with the construction of Brazil's Perimetral Norte Highway, which cut through two hundred kilometers of their territory. The road workers introduced contagious diseases and prostitution. By 1975, four Yanomami villages on the Ajarani River had lost 22 percent of their people. By 1978, four villages on the Catrimani

River had lost 50 percent of their people. In one Yanomami community in the Serra de Surucucus, 68 percent of the Yanomami died virtually overnight.

A second wave of death began gathering with the discovery of gold on Yanomami lands. In August 1987, thousands of *garimpeiros*, the go-for-broke prospectors and roustabouts that chase every real or imagined mother lode, were pouring into the area from every corner of Brazil. By the next spring, the miners had built more than eighty illegal airstrips. During the first half of 1989, fifty thousand miners had invaded Yanomami territory. "In a matter of months," noted a report to the Brazilian government by Alcida Ramos, "major rivers became unusable. Mercury pollution [from refining gold] and silting drastically affected the entire course of the Macajaí River; poisoned with mercury and oil, the Uraricoera, Catrimani, and Couto de Magalhaẽs rivers stopped yielding fish. . . . In the first nine months of 1990, more than fifteen hundred Yanomami died, mostly from diseases, but many had been shot."

The Threat to African Pygmies

Logging is threatening forest peoples' ways of life. In Zaire, an elderly Baka Pygmy man says, "We do not kill so much game now as we used to. We wonder what our children will eat in the future." A Baka woman agrees, "In the past we made *molongo* [family treks of several months into the forest] but we no longer do it. We killed a lot of meat with spears. Today there are not many true Baka who kill a lot of meat. We kill a little, a little." And many of the Aka Pygmies of Zaire who have been drawn out of the forest to work in the logging camps have fallen prey to disease and alcoholism.

In Rwanda, the forests inhabited by the Twa Pygmies are almost entirely gone. Some of the Rwanda Twa have only recently been deprived of their hunting and gathering existence by the deforestation of the country, while others have for generations lived outside the forest as an oppressed minority forced to eke out an existence by begging and squatting.

Virginia Luling and Damien Lewis, *Multinational Monitor*, September 1992.

"When I was young, we weren't suffering like this," Davi Kopenawa Yanomami told me in Rio de Janeiro in the spring of 1992. "When I was a boy, life was good. We weren't dying like we are today. Since the white people have come to our country, they have let us live in hunger, they let us suffer. I am talking like this so you can feel it inside your heart."

As one of the very few Yanomami who speak Portuguese, Davi has often been the sole emissary between his people and the outside world. When we've seen each other in Rio de Janeiro and Kari-Oca, Brazil, and later at the United Nations in New York, he's always been very warm and direct. And every time we've sat together to talk, I've had the feeling of being in the presence of a person from another time and place. This sensation must arise in part because he comes from a place in the forest I will probably never see and from a childhood thousands of years removed from my own. But the pain in his voice is familiar. . . .

"The Yanomami are sick. We are dying," Davi said. "We don't know how to take care of the disease the gold miners brought to our land. It is the government's fault. It didn't think about us when it allowed the gold seekers to come in. We want to ask the world to help us pressure the Brazilian government into taking care of the indigenous peoples all over Brazil. We are asking them to send us doctors to exterminate this disease. I don't want my people to die."

Visions

For many years Claudia Andujar of São Paulo, Brazil, has worked with Davi and other Yanomami to protect their lands. "All the people who have studied the Yanomami and know them have different visions of the Yanomami," she once said. "I see them as intellectuals, as people who think about and understand the world in a global way. It is a vision as old as humanity, but we have lost it."

From my brief but memorable encounters with Davi Yanomami, I tend to share Andujar's view. "I need the help of your people to cure this sickness," Davi told me. "But concerning nature, I need to help your people. The Americans, the English, the Japanese—we want to teach their children and grandchildren not to destroy the earth anymore. The situation isn't dangerous just for the Yanomamis but for everyone. We all live on the same planet."

And then, looking straight into my face, he said, as if he were talking both to me very personally and to my people, "I'm with you in this fight. I am not going to run away. But I am not going to be quiet. Whenever you need me, I'll help you, because the Yanomami need to fight in order that our brothers will not suffer.

"We don't have jobs and money. And we don't want these things from other cultures. Indians like us live in another world. And we want to remain Indians." Davi paused to make sure I understood. "The mountains, the rain, the wind, the moon, the stars, the sun—we need these to keep living. We want to remain in our culture, in our way of thinking. I want to be in my forest, listening to the birds, the thunders—breathing the pure air."

VIEWPOINT 2

"Indigenous groups . . . are . . . adapting traditional resource management arrangements to the modern setting."

Indigenous Peoples Are Adapting to the Modern World

Alan Thein Durning

Although indigenous peoples face many problems, they can adapt their traditional ways of life to fit within modern civilization, Alan Thein Durning asserts in the following viewpoint. Durning argues that some native groups have secured the legal right to practice traditional fishing, hunting, and agricultural practices—an example other indigenous cultures can follow. Utilizing native knowledge to create products and medicines for the modern market could provide indigenous tribes with a steady cash income, Durning maintains. A senior researcher at Worldwatch Institute, Durning is the author of *How Much Is Enough? The Consumer Society and the Future of the Earth.*

As you read, consider the following questions:

1. In Durning's opinion, why do Western scientists often overlook indigenous knowledge?
2. According to the author, what three conditions are necessary for traditional stewardship to survive?
3. What two recent developments can assist indigenous communities, in Durning's view?

Excerpted from Alan Thein Durning, "Guardians of the Land: Indigenous Peoples and the Health of the Earth," *Wordwatch Paper 112*, December 1992. Copyright 1992 by The Worldwatch Institute. Reprinted with permission.

Sustainable use of local resources is simple self-preservation for people whose way of life is tied to the fertility and natural abundance of the land. Any community that knows its children and grandchildren will live exactly where it does is more apt to take a long view than a community without attachments to local places. Moreover, native peoples frequently aim to preserve not just a standard of living but a way of life rooted in the uniqueness of a local place. Colombian anthropologist Martin von Hildebrand notes, "The Indians often tell me that the difference between a colonist [a non-Indian settler] and an Indian is that the colonist wants to leave money for his children and that the Indians want to leave forests for their children."

Nature

In most native cosmologies, furthermore, nature is more than a storehouse of resources. Says Salvador Raín of the Mapuche of Chile, "Our relationship with our god Guinechén is expressed through the land. It is where we develop our culture and bury our dead." Like the Mapuche, many indigenous cultures view nature as inherently valuable, revere it as an embodiment of the divine, or honor it as the home of ancestors' spirits. Their religious beliefs are manifested in the rituals of offering, thanksgiving, and spiritual cleansing so commonly required of those who take fish, game, or trees from their domain. Amid the endless variety of indigenous belief, there is striking unity on the sacredness of ecological systems.

Indigenous peoples' veneration for the natural world has its parallel in their peerless ecological knowledge. Most forest-dwelling tribes are masters of botany. The Shuar people of Ecuador's Amazonian lowlands, for example, use 800 species of plants for medicine, food, animal fodder, fuel, construction, fishing, and hunting supplies. Traditional healers in southeast Asia rely on as many as 6,500 medicinal plants, and shifting cultivators throughout the tropics frequently sow more than 100 crops in their forest farms.

Native peoples commonly know as much about ecological processes that affect the availability of natural resources as they do about those resources' diverse uses. South Pacific islanders can predict to the day and hour the beginning of the annual spawning runs of many fish. Whaling peoples of northern Canada have proved to skeptical western marine biologists that bowhead whales migrate under pack ice. Coastal aborigines in Australia distinguish between 80 different tidal conditions.

Specialists trained in western science often fail to recognize indigenous knowledge because of the cultural and religious forms in which indigenous peoples record and transmit it. Ways of life that developed over scores of generations could only thrive by encoding ecological sustainability into the body of practice, myth, and taboo that passes from parent to child. . . .

The quality of native stewardship is evident in comparisons of the ecological condition of indigenous lands with that of neighboring lands managed by others. The island of New Guinea—divided politically between the Indonesian province of Irian Jaya and the independent country of Papua New Guinea—has more distinct cultures than any comparable land area on earth, with more than one-sixth of the world's languages on an island the size of Turkey. Of all the world's major tribal areas, New Guinea was the last to be reached by outsiders, who only first made contact with inland peoples in the early 1960s. Most of Papua New Guinea is controlled by leaders of local tribes under customary land rights, while in Irian Jaya, land management decisions are made exclusively by the state. The consequences for local peoples, natural habitats, and social equity are clear: In Papua New Guinea, although indigenous groups have sold resources from their homelands to loggers and miners, they have done so slowly, without devastating their own cultures, and have received some of the profits. In Irian Jaya, indigenous people have had their resources stolen, their cultures devastated, and their subsistence economies gutted. Indonesia has designated three-fourths of the province's forests for timber concessionaires.

Vulnerability

Ingenious as customary stewardship arrangements are, they are also vulnerable. When pressures come to bear from the cash economy, powerful modern technologies, and encroaching populations—or, occasionally, from their own growing numbers—native stewards are likely to find their traditional approaches to management collapsing. The results for nature are catastrophic. In these circumstances, indigenous peoples—like anyone else—are prone to overuse resources, overhunt game, and sell off timber and minerals to pay for consumer goods. Indeed, Alaskan natives have scarred their lands with some of the worst clear-cuts in the United States, and members of the Philippine Manobo tribe now hunt for wild pigs by packing explosives into overripe fruit. Similarly, when new roads in West Kalimantan, Indonesia, made ironwood logging profitable in the late eighties, young men of the Galik tribe bought or borrowed chain saws and rapidly cleared the forest of the ancient trees.

What are the conditions in which traditional systems of ecological management can persist in the modern world? Based on the diverse experience of indigenous peoples, three necessary conditions stand out. First, traditional stewardship's persistence depends on indigenous peoples' having secure rights to their subsistence base—rights that are not only recognized but enforced by the state and, ideally, backed by international law. Latin American tribes such as the Shuar of Ecuador, when threatened with losing their land, have cleared their own forests and taken up cattle ranching, because in Latin America these actions prove ownership. Had Ecuador defended the Shuar's land rights, the ranching would have been unnecessary.

Second, indigenous ecological stewardship can survive the onslaught of the outside world if indigenous peoples are organized politically and the states in which they reside allow democratic initiatives. The Khant and Mansi peoples of Siberia, like most indigenous people in the former Soviet Union, were nominally autonomous in their customary territories under Soviet law, but political repression precluded the organized defense of that terrain until the end of the eighties. Since then, the peoples of Siberia have begun organizing themselves to turn paper rights into real local control. In neighboring China, in contrast, indigenous homelands remain nothing more than legal fictions because the state crushes all representative organizations.

Third, if they are to surmount the obstacles of the outside world, indigenous communities need access to information, support, and advice from friendly sources. The tribal people of Papua New Guinea know much about their local environments, for example, but they know little about the impacts of large-scale logging and mining. Foreign and domestic investors have often played on this ignorance, assuring remote groups that no lasting harm would result from leasing parts of their land to resource extractors. If the forest peoples of Papua New Guinea could learn from the experience of threatened peoples elsewhere—through supportive organizations and indigenous peoples' federations—they might be more careful.

Maintaining Traditions

A handful of peoples around the world have succeeded in satisfying all three of these conditions. One example is the salmon-based cultures of the Pacific Northwest. In treaty negotiations a century ago, the U.S. government promised these cultures permanent access to their customary fishing grounds both on and off reservations, in exchange for territorial concessions. But starting early in this century, non-Indian fishers began to take most of the catch, leaving little for Indians. The native fishing industry dwindled, and by mid-century had almost died, until the Indians organized themselves to demand their rights. Eventually, in a series of landmark legal rulings in the seventies, U.S. courts interpreted the treaties as reserving half of all disputed fish for Indians.

Their rights secured, the tribes have once again become accomplished fishery managers—rejuvenating their traditional reverence for salmon and training themselves in modern approaches with the help of supportive non-Indians. As stipulated under the court rulings, state and federal fisheries regulators have agreed with qualified tribes to manage fish runs jointly. Today, the Lummi, Tulalip, Muckleshoot, and other Northwestern tribes are managing the salmon runs that nourished their ancestors.

A similar case can be found in Namibia, where most of the San—after a century in which their population declined by 80 percent and their land base shrank by 85 percent—are now day laborers on cash-crop plantations. In the eighties, however, some 48 bands of San, totalling about 2,500 individuals, organized themselves to return to the desert homes they had tenuous rights over. There they have created a modified version of their ancient hunting and gathering economy. With the help of anthropologists, they have added livestock and drip-irrigated gardens to the daily foraging trips of their forebears, fashioning a way of life that is both traditional and modern.

Perhaps because natural resource rights are best recognized in the Americas, indigenous groups there are furthest advanced in adapting traditional resource management arrangements to the modern setting. In northern Canada, the Inuvialuit people have created management plans for grizzly and polar bears and for beluga whales. In southern Mexico, the Chinantec Indians are gradually developing their own blend of timber cutting, furniture making, butterfly farming, and forest preservation in their retreat in the Juarez mountains. The Miskito Indians of Nicaragua's Atlantic Coast, meanwhile, are forming local management groups to police the use of forests, wetlands, and reefs in the extensive Miskito Coast Protected Area they helped create in 1991.

Cooperation

As they struggle to adapt their natural resource stewardship to modern pressures, indigenous peoples are beginning to pool their expertise. The Native Fish and Wildlife Service in Colorado, formed by a coalition of North American tribes, serves as an information clearinghouse on sustainable management. The Kuna of Panama—whose tribal regulations on hunting turtles and game, catching lobsters, and felling trees fill thick volumes—have convened international conferences on forest and fisheries management. The Inuit Circumpolar Conference, representing Inuit peoples from Canada, Greenland, Russia, and the United States, has developed an Inuit Regional Conservation Strategy that includes tight controls on wildlife harvesting and resource extraction, and collaborative arrangements for sharing ecological knowledge. Such instances are still exceptional, but they blaze a trail for indigenous peoples everywhere.

Where the basic conditions of land or water rights, political organization, and access to information are satisfied, two additional and relatively recent developments—new approaches to trade and intellectual property rights—could further assist indigenous communities' efforts to sustain their systems of stewardship against outside forces. These developments also promise native peoples greater control over their interaction with the money economy.

Alternative traders, organizations committed to cultural survival and environmental sustainability, now market millions of dollars' worth of indigenous peoples' products in industrial countries. The Mixe Indians of southern Mexico, for example, sell organic coffee to U.S. consumers through the Texas-based alternative trade organization Pueblo to People. The Kayapó sell Brazil-nut oil for use in hair conditioners to the U.K.-based alternative trader The Body Shop.

Protecting Their Land

Land rights have been the focus of most indigenous campaigning. A series of seminal protests—the 1992 uprising in Ecuador, the 1990 march for 'Land and Dignity' in Bolivia—secured presidential decrees guaranteeing huge tracts. . . .

Throughout the Amazon Basin indigenous guards now patrol the frontiers of huge tracts on foot, by canoe or in four-wheel-drive vehicles. Many carry shortwave radios linking them to the nearest village where faxes, computer networks and telephones can alert local and international support networks.

Phillip Wearne, *The New Internationalist*, June 1994.

By eliminating links from the merchandising chain, Pueblo to People, The Body Shop, and other alternative traders keep more of the product value flowing back to indigenous producers. The potential for alternative trade to grow is enormous, given the growing purchasing power of environmentally conscious consumers and the abundance of plant products hidden in indigenous lands. Mexico's forests hold an estimated 3,000 useful substances known only to Indians. Among the Quechua of lowland Ecuador, each hectare of forest yields fruits, medicinal plants, and other products worth $1,150 per year in Ecuadorean urban markets. While trade will never make indigenous peoples rich, it can provide them with a modest cash income to supplement their subsistence. Admittedly, harvesting wild products for national and international markets has its risks: it can fuel overexploitation of resources and create schisms within communities. Still, in a world where money is power, no group can survive long without some source of revenue.

Medicinal Patents

A newer route to assuring indigenous peoples a basic income traverses the legal terrain of intellectual property rights—proprietary rights to ideas, designs, or information most commonly typified by patents and copyrights. Indigenous peoples have painstakingly studied their environments, and their knowledge has aided billions of people elsewhere, when, for example, their medicinal plants became the sources of life-saving drugs. Hundreds of years ago the highland Quechua of Peru revealed to Europeans the anti-malarial medicine quinine; the lowland Indians of the Amazon showed them how to make the emetic ipecac; and the peoples of the Guyanas instructed them on extracting from plants the muscle-relaxant curare, which was used in abdominal surgery for a century.

Yet indigenous peoples have rarely received anything of commensurate value in return; indeed, they have sometimes been annihilated for their efforts. One fourth of the coal the Soviet Union dug up during its seven-decade history came from the lands of the Shorish people of Western Siberia. The destruction that mining unleashed after the Shors disclosed the locations of the minerals has whittled them down to a few hundred survivors. In Guyana, likewise, the Macushi tribe revealed the ingredients of their blow-dart poison to English naturalist Charles Waterton in 1812. Scientists used that recipe to develop curare. The tribe—dispossessed, uprooted, and alienated from its culture in the intervening period—now lives in misery, not even remembering how to make blow guns.

The potential payoff for establishing indigenous peoples' rights to their knowledge is enormous, as the case of prescription medicines illustrates. One-fourth of prescription drugs dispensed by U.S. pharmacies are derived from plants. Of those plant-derived active ingredients, approximately three-fourths have similar uses in traditional, herbal medicine, according to pharmacologist Norman R. Farnsworth of the University of Illinois in Chicago. Many were developed by following the lead of indigenous healers. Annual sales of plant-derived drugs in the United States alone total $8 billion, so even if indigenous peoples earned only a fraction of one percent in royalties for their contributions to drug development, the cash flow would be substantial.

With the explosive growth in biotechnology since 1980, the demand for new genetic material is burgeoning. Many of the world's genes are in the millions of species in the endangered places known only to endangered peoples. Indeed, some indigenous leaders think of the rush to codify and exploit indigenous knowledge of biological diversity as the latest in the long history of resource grabs perpetrated against them. "Today," says Adrian Esquina Lisco, spiritual chief of the National Association of Indigenous Peoples of El Salvador, "the white world wants to understand the native cultures and extract those fragments of wisdom which extend its own dominion." Still, supporters of indigenous peoples are developing legal strategies to turn the gene trade to native advantage by demanding recognition that indigenous communities possess intellectual property rights as valid as those of other inventors and discoverers.

Native peoples' cultural ties to their local environments predispose them to guard and conserve the flora and fauna of their ancestral homes, but they need rights to their subsistence base, a degree of political organization, and support from allied segments of the world beyond their borders to translate that cultural predisposition into sustainable development. The world's indigenous peoples expend much of their energy simply trying to secure the first of those conditions: resource rights. To date, their successes have been few. But there is reason to expect greater success in achieving this first condition as advances continue in the second and third conditions—indigenous political mobilization and support from nonindigenous people.

"The loss of life's diversity endangers not just the body but the spirit."

Humans Are Innately Connected to Nature

Edward O. Wilson

In 1984, Edward O. Wilson published *Biophilia* (literally, "love of life"), in which he put forth his theory that humans are inherently dependent on nature not just physically, but emotionally and mentally as well. In his more recent book *The Diversity of Life*, from which the following viewpoint is excerpted, Wilson contends that humanity's increasing alienation from the natural world harms the human spirit. Loss of wilderness and species extinction affect the human spirit in ways that are barely beginning to be understood, Wilson maintains. The author of numerous books and articles, Wilson has twice won the Pulitzer Prize.

As you read, consider the following questions:

1. In Wilson's opinion, what is the primary cause of humans' lack of knowledge about their true nature?
2. What human behaviors does the author believe are examples of biophilia?
3. What ethical imperative should be followed in wildlife conservation, in Wilson's view?

Reprinted by permission of the publishers from *The Diversity of Life* by Edward O. Wilson, Cambridge, Mass.: The Belknap Press of Harvard University Press, ©1992 by Edward O. Wilson.

It is . . . possible for some to dream that people will go on living comfortably in a biologically impoverished world. They suppose that a prosthetic environment is within the power of technology, that human life can still flourish in a completely humanized world, where medicines would all be synthesized from chemicals off the shelf, food grown from a few dozen domestic crop species, the atmosphere and climate regulated by computer-driven fusion energy, and the earth made over until it becomes a literal spaceship rather than a metaphorical one, with people reading displays and touching buttons on the bridge. Such is the terminus of the philosophy of exemptionalism: do not weep for the past, humanity is a new order of life, let species die if they block progress, scientific and technological genius will find another way. Look up and see the stars awaiting us.

But consider: human advance is determined not by reason alone but by emotions peculiar to our species, aided and tempered by reason. What makes us people and not computers is emotion. We have little grasp of our true nature, of what it is to be human and therefore where our descendants might someday wish we had directed Spaceship Earth. Our troubles, as Vercors said in *You Shall Know Them*, arise from the fact that we do not know what we are and cannot agree on what we want to be. The primary cause of this intellectual failure is ignorance of our origins. We did not arrive on this planet as aliens. Humanity is part of nature, a species that evolved among other species. The more closely we identify ourselves with the rest of life, the more quickly we will be able to discover the sources of human sensibility and acquire the knowledge on which an enduring ethic, a sense of preferred direction, can be built.

The human heritage does not go back only for the conventionally recognized 8,000 years or so of recorded history, but for at least 2 million years, to the appearance of the first "true" human beings, the earliest species composing the genus *Homo*. Across thousands of generations, the emergence of culture must have been profoundly influenced by simultaneous events in genetic evolution, especially those occurring in the anatomy and physiology of the brain. Conversely, genetic evolution must have been guided forcefully by the kinds of selection rising within culture.

Humans and Nature

Only in the last moment of human history has the delusion arisen that people can flourish apart from the rest of the living world. Preliterate societies were in intimate contact with a bewildering array of life forms. Their minds could only partly adapt to that challenge. But they struggled to understand the most relevant parts, aware that the right responses gave life and fulfillment, the wrong ones sickness, hunger, and death. The imprint of that effort cannot have been erased in a few generations of urban existence. I suggest that it is to be found among the particularities of human nature, among which are these:

• People acquire phobias, abrupt and intractable aversions, to the objects and circumstances that threaten humanity in natural environments: heights, closed spaces, open spaces, running water, wolves, spiders, snakes. They rarely form phobias to the recently invented contrivances that are far more dangerous, such as guns, knives, automobiles, and electric sockets.

Reprinted with special permission of King Features Syndicate.

• People are both repelled and fascinated by snakes, even when they have never seen one in nature. In most cultures the serpent is the dominant wild animal of mythical and religious symbolism. Manhattanites dream of them with the same frequency as Zulus. This response appears to be Darwinian in origin. Poisonous snakes have been an important cause of mortality almost everywhere, from Finland to Tasmania, Canada to Patagonia; an untutored alertness in their presence saves lives. We note a kindred response in many primates, including Old World monkeys and chimpanzees: the animals pull back, alert others, watch closely, and follow each potentially dangerous snake until it moves away. For human beings, in a larger metaphorical sense, the mythic, transformed serpent has come to possess both constructive and destructive powers: Ashtoreth of the Canaanites, the demons Fu-Hsi and Nu-kua of the Han Chinese, Mudamma and Manasa of Hindu India, the triple-headed giant Nehebkau of the ancient Egyptians, the serpent of Genesis conferring knowledge and death, and, among the Aztecs, Cihuacoatl, goddess of childbirth and mother of the human race, the rain god Tlaloc, and Quetzalcoatl, the plumed serpent with a human head who reigned as lord of the morning and evening star. Ophidian power spills over into modern life: two serpents entwine the caduceus, first the winged staff of Mercury as messenger of the gods, then the safe-conduct pass of ambassadors and heralds, and today the universal emblem of the medical profession.

• The favored living place of most peoples is a prominence near water from which parkland can be viewed. On such heights are found the abodes of the powerful and rich, tombs of the great, temples, parliaments, and monuments commemorating tribal glory. The location is today an aesthetic choice and, by the implied freedom to settle there, a symbol of status. In ancient, more practical times the topography provided a place to retreat and a sweeping prospect from which to spot the distant approach of storms and enemy forces. Every animal species selects a habitat in which its members gain a favorable mix of security and food. For most of deep history, human beings lived in tropical and subtropical savanna in East Africa, open country sprinkled with streams and lakes, trees and copses. In similar topography modern peoples choose their residences and design their parks and gardens, if given a free choice. They simulate neither dense jungles, toward which gibbons are drawn, nor dry grasslands, preferred by hamadryas baboons. In their gardens they plant trees that resemble the acacias, sterculias, and other native trees of the African savannas. The ideal tree crown sought is consistently wider than tall, with spreading lowermost branches close enough to the ground to touch and climb, clothed with compound or needle-shaped leaves.

• Given the means and sufficient leisure, a large portion of the populace backpacks, hunts, fishes, birdwatches, and gardens. In the United States and Canada more people visit zoos and aquariums than attend all professional athletic events combined. They crowd the national parks to view natural landscapes, looking from the tops of prominences out across rugged terrain for a glimpse of tumbling water and animals living free. They travel long distances to stroll along the seashore, for reasons they can't put into words.

Biophilia

These are examples of what I have called *biophilia*, the connections that human beings subconsciously seek with the rest of life. To biophilia can be added the idea of wilderness, all the land and communities of plants and animals still unsullied by human occupation. Into wilderness people travel in search of new life and wonder, and from wilderness they return to the parts of the earth that have been humanized and made physically secure. Wilderness settles peace on the soul because it needs no help; it is beyond human contrivance. Wilderness is a metaphor of unlimited opportunity, rising from the tribal memory of a time when humanity spread across the world, valley to valley, island to island, godstruck, firm in the belief that virgin land went on forever past the horizon.

I cite these common preferences of mind not as proof of an innate human nature but rather to suggest that we think more carefully and turn philosophy to the central questions of human origins in the wild environment. We do not understand ourselves yet and descend farther from heaven's air if we forget how much the natural world means to us. Signals abound that the loss of life's diversity endangers not just the body but the spirit. If that much is true, the changes occurring now will visit harm on all generations to come.

The ethical imperative should therefore be, first of all, prudence. We should judge every scrap of biodiversity as priceless while we learn to use it and come to understand what it means to humanity. We should not knowingly allow any species or race to go extinct. And let us go beyond mere salvage to begin the restoration of natural environments, in order to enlarge wild populations and stanch the hemorrhaging of biological wealth. There can be no purpose more enspiriting than to begin the age of restoration, reweaving the wondrous diversity of life that still surrounds us.

The evidence of swift environmental change calls for an ethic uncoupled from other systems of belief. Those committed by religion to believe that life was put on earth in one divine stroke will recognize that we are destroying the Creation, and those who perceive biodiversity to be the product of blind evolution will agree. Across the other great philosophical divide, it does not matter whether species have independent rights or, conversely, that moral reasoning is uniquely a human concern. Defenders of both premises seem destined to gravitate toward the same position on conservation.

An Environmental Ethic

The stewardship of environment is a domain on the near side of metaphysics where all reflective persons can surely find common ground. For what, in the final analysis, is morality but the command of conscience seasoned by a rational examination of consequences? And what is a fundamental precept but one that serves all generations? An enduring environmental ethic will aim to preserve not only the health and freedom of our species, but access to the world in which the human spirit was born.

VIEWPOINT 4

> "*Most . . . Westerners . . . grow up dependent on human artifacts, live surrounded by them, and learn a great deal about them.*"

Humans Are Not Innately Connected to Nature

Jared Diamond

Many scientists and environmentalists support the biophilia hypothesis, which proposes that all humans have an inherent connection to nature. However, Jared Diamond argues in the following viewpoint, New Guineans and other peoples who have only recently been exposed to modern civilization do not exhibit the characteristics of biophilia. Such people have little emotional affinity with animals, are often needlessly cruel to wildlife, and do not evidence an innate fear of snakes or love of the forest, Diamond argues. Based on these observations, Diamond concludes that biophilia is not a universal human trait. A professor of physiology at the University of California Medical School, Diamond is the author of *The Third Chimpanzee*.

As you read, consider the following questions:
1. According to Diamond, do any Stone Age peoples of the modern world keep pets?
2. Why do modern urbanized people tend to fear all snakes and spiders, in Diamond's opinion?
3. According to the author, in what ways do New Guineans who practice traditional lifestyles show an affinity with nature?

Excerpted from "New Guineans and Their Natural World" by Jared Diamond, in *The Biophilia Hypothesis*, edited by Stephen R. Kellert and Edward O. Wilson. Copyright ©1993 by Island Press. Reprinted by permission of Alexander Hoyt Associates for Island Press.

There are several reasons why New Guinea (together with other Pacific islands) seems a good place for gathering empirical data relevant to testing the biophilia hypothesis. First, all humans were wholly dependent on stone technology until about 5000 B.C., and only since then has one human group after another begun using metals. New Guineans were among the last to switch to metal: all New Guineans depended on stone technology until European colonization began in the nineteenth century. Of the groups of New Guineans with whom I have worked, most remained uncontacted by the outside world and still used stone tools until the 1950s, and some continued in that way up to the 1980s. Thus New Guineans furnish some of our best surviving models of the human conditions that have prevailed for millennia. . . .

With respect to habitat diversity, most of New Guinea is covered with closed forest. Southern New Guinea has three areas of savanna woodland similar to that of northern Australia. Other natural habitats include the seacoast, rivers, and alpine grassland, while large areas of the highlands are clothed in anthropogenic grassland. The closed forest itself is very varied, ranging from lowland tropical rain forest and swamp forest to montane oak forest and montane beech forest. Some of the largest areas of intact tropical forest in the world are to be found in New Guinea. For example, I have frequently had the experience of flying for 200 miles by small aircraft over the New Guinea lowland forest and seeing no signs of human occupation except for one or two huts of nomads. I should add that I am puzzled, in other discussions of the biophilia hypothesis, by what seems to me an exaggerated focus on savanna habitats as a postulated influence on innate human responses. Humans spread out of Africa's savannas at least 1 million years ago. We have had plenty of time since then—tens of thousands of generations—to replace any original innate responses to savanna with innate responses to the new habitats encountered. The earliest attested behaviorally modern *Homo sapiens* were the ancestors of aboriginal Australians, who between 40,000 and 30,000 years ago occupied Australia's entire span of habitats—from desert, the seacoast, tropical rain forest, dry forest, mulga, Mediterranean scrub, and heath to cold temperate rain forest and periglacial grassland. . . .

Attitudes Toward Animals

Do [New Guineans] exhibit any positive emotional responses to animals as living creatures—responses such as love, reverence, fondness, concern, or sympathy? New Guineans certainly are capable of positive responses to at least one species of domestic animal, the pig, which serves as a major status symbol and with which they live on intimate terms. Young pigs often sleep in the same hut with their human owners, and New Guinea women sometimes nurse a piglet at one breast while nursing their own infant at the other. I have seen New Guineans become upset when their pig dies and get angry when they think somebody else is responsible for the death of their pig.

It is rare, however, to see corresponding signs of New Guineans recognizing individual wild animals as living creatures to which one can form a bond. Wild animals are only rarely kept as pets in New Guinean or Melanesian villages. The only examples I have encountered are a couple of pet hornbills, several cases of imperial pigeons, and one case of a young tree kangaroo. (I do not count as pets the possums that one sometimes sees trussed up in villages, being kept alive after capture until their intended sacrifice for food.) This infrequency of village pets is surprising to Westerners, because so many New Guinea wild mammal and bird species tame well and make cute, responsive pets much beloved by Western expatriates living in New Guinea. New Guineans' disinterest in pets can certainly not be generalized to other Stone Age peoples of the modern world: I saw numerous pets in a single village of recently contacted Ishcanahua Amerindians whom I visited on a tributary of the Peruvian Amazon.

It is not only that New Guineans fail to cultivate friendly responses from wild animals. They also seem not to take account of the fact that these are living creatures capable of feeling pain. For example, when a wild animal is captured in the forest early in the day and is to be transported alive for the rest of the day so that it can be killed and eaten fresh in the village that evening, the animal's legs may be broken to prevent it from escaping. One painful experience stands out vividly in my mind. On one occasion it turned out that the man whom I had hired as my guide for the day in the Aru Islands was a bird trapper retrieving wild cockatoos that he had snared. The cockatoos were not intended for consumption but for sale to a trader who would eventually sell them for illegal export to foreign fanciers of cockatoos as pets. In this case the man immobilized the cockatoos for transport during the day by the barbarous method of bending their wings behind their backs and then tying several of their primary feathers together in a knot. Since the man was armed with a bush knife and I was alone with him in a remote area of forest, I was helpless to interfere.

Indifference

As another example of indifference to needless suffering of wild animals, I recall an occasion on my first New Guinea expedition when I heard squealing and shrieking ahead of me as I returned to my campsite. At the camp I found several New Guineans holding a large fruit bat of a species whose long and slender wingbones are used as nose decorations. The men wanted to cut these bones out of the bat's wing, but they did not bother to kill the bat before doing so. Instead, the two men spread out the bat's wings, another tied up the bat's mouth with a vine to prevent it from biting, and the other man then proceeded to dissect and scrape through the joints and muscles of the live bat so as to extract the bones.

In another case, I found men intentionally inflicting pain on captured live bats for no other reason than amusement at the reactions of the tortured animals. The men had tied twenty-six small *Syconycteris* blossom bats to strings. They lowered one bat after another until it touched the red-hot embers of a fire, causing the bat to writhe and squeal in pain. The men raised the bat, lowered it again for another touch to the red-hot embers, repeated this process until it was dead, and then went on to the next bat, finding the whole proceedings funny. . . .

Biophobia

If there is thus little evidence of love or other tender feelings for individual wild animals, is there any evidence of fear of animals, such as snakes and spiders, to which people in industrialized societies are often described as possessing an innate fear (biophobia)? If there is any single place in the world where we might expect to find an innate fear of snakes among native peoples, it would be New Guinea, where one-third or more of the snake species are poisonous and certain nonpoisonous constrictor snakes are sufficiently big to be dangerous. One of the few well-attested examples of a large snake actually killing and eating a human involves a reticulated python consuming a fourteen-year-old boy on an Indonesian island.

In one case I did observe fear of snakes on the part of New Guineans: two men returned from a morning's pursuit of birds in the forest to report that they had encountered a very large python, that they were afraid of it, and that they did not want to return on that trail again. This can hardly be considered an example of irrational innate fear; instead, it is a completely appropriate learned response to a dangerous animal. New Guineans certainly possess no generalized fear of snakes and spiders as many Westerners do. They are well aware which species are poisonous and which are not. When a New Guinean and I encounter a snake in the forest together, the New Guinean simply explains to me matter-of-factly whether that particular snake is dangerous. Nonpoisonous snakes are routinely captured by children and women, as well as by men, for eating. Children capture large spiders, singe off the legs and hairs, and eat the bodies. Asked whether they have a generalized fear of snakes, New Guineans laugh in scorn and say that that is a reaction for ignorant white men too stupid to distinguish poisonous from nonpoisonous snakes. . . .

What about the supposedly widespread, innate fear of snakes reported in cross-cultural studies—of people as seemingly diverse as Americans, Europeans, Japanese, white Australians, and Argentineans? These studies actually refer to only a tiny slice of human cultural diversity—a slice composed of modern industrialized metal-using peoples living in centralized political states. Such people have good reason to fear snakes. Distinguishing poisonous from nonpoisonous snakes can be difficult, especially since certain harmless species have evolved into excellent mimics of poisonous ones. . . .

Modern Cultures and Nature

Despite . . . assertions of an especially refined appreciation for nature in the United States and Japan, research has revealed only limited concern for the natural world among the general public in both countries. Citizens of the United States and Japan typically expressed strong interest in nature only in relation to a small number of species and landscapes characterized by especially prominent aesthetic, cultural, and historic features. Furthermore, most Americans and Japanese expressed strong inclinations to exploit nature for various practical purposes despite the likelihood of inflicting considerable environmental damage. . . .

Japanese appreciation of nature was especially marked by a restricted focus on a small number of species and natural objects—often admired in a context emphasizing control, manipulation, and contrivance. This affinity for nature was typically an idealistic rendering of valued aspects of the natural environment, usually lacking an ecological or ethical orientation.

Stephen R. Kellert, *The Biophilia Hypothesis*, edited by Stephen R. Kellert and Edward O. Wilson, 1993.

Foraging peoples who eat snakes have to learn these distinctions. However, snakes do not make a major contribution to the diet of any foraging people. Over the past 10,000 years, as domesticated animals and plants have increasingly replaced wild foods in the diets of most peoples, snakes must have been among the first wild food items to be dropped from the diet: too much danger, too much specialized knowledge required, too little payoff. In nonforaging societies it doesn't make sense for people to waste time learning to distinguish snake species; it's better to learn as babies from their parents' frightened responses a generalized fear of snakes, to be passed on in turn to their children. The same reasoning applies to spider species: worth distinguishing and selectively eating, for New Guinea children; not worth distinguishing, and only worth generically learning to fear, for children in nonforaging societies. Probably it was not until recent millennia, as snakes and spiders lost their traditional value as minor food items, that most human societies developed a learned generalized biophobia of snakes and spiders....

The Forest

New Guineans' knowledge of local plants and animals is certainly a learned product of their experience. At least in large part, their attitudes toward their natural environment are also a learned product of experience. Let me cite two examples.

First, Indonesians and New Guineans from forested areas are at ease in the forest and live contentedly inside it for weeks at a time. Indonesians from deforested areas are afraid of the forest and reluctant to enter it. When I was carrying out a national park survey of the Fakfak Mountains of Indonesian New Guinea in 1981 and 1983, a wonderful Indonesian forester named Lefan explained to me how he and other members of his department felt about the forest. Lefan and some of his close colleagues had grown up at home in the forest, enjoyed working in it on timber surveys, and were accustomed to spending long periods in it dozens of kilometers from the coast. Lefan's boss, however, was an Indonesian who came from a deforested area and preferred to stay at his desk. When I asked Lefan whether his boss entered the forest at all, Lefan replied (in Indonesian): *"Ya, masuk ke hutan. Masuk seratus meter, dan kembali."* ("Sure, he goes into the forest. He goes in 100 meters, then he comes out again.")

My second example concerns young New Guineans whose parents used Stone Age technology and exploited the forest, but who themselves go to school and eventually move to a town in search of work. Naturally, these young New Guineans have no opportunity to acquire detailed knowledge of the forest. Perhaps surprisingly, they exhibit little interest in it either. The national parks and zoos that the governments of Papua New Guinea and Indonesia have established near urban areas, in hopes of stimulating interest by urbanized peoples in their country's natural heritage, elicit disappointingly little interest, except by people using the open spaces for picnicking and Sunday outings. Their fear of the forest, as well as their negligible interest in the natural heritage with which their ancestors lived so intimately for tens of thousands of years, are a big problem for the governments of Indonesia and Papua New Guinea today in their effort to develop indigenous support for conservation.

In short, New Guineans and other Pacific islanders still practicing traditional lifestyles possess a deep and detailed knowledge of wild plant and animal species, a dependence on their natural environment for their economy, and a use of wild species for their decorations and status symbols and myths—to a degree, indeed, that is difficult for those of us reared in urbanized Western societies to grasp. If biophilia is defined as human affinity—regardless of whether learned or innate—for other species, New Guineans serve as textbook examples.

If, however, biophilia is specifically defined as an innate affinity, it is not presently clear to me what might constitute evidence for such a genetic basis. Yes, New Guineans with traditional lifestyles grow up to depend on other species, live surrounded by them, and learn a great deal about them. Most of us Westerners instead grow up dependent on human artifacts, live surrounded by them, and learn a great deal about them. To a striking degree, the human brain is a generalized data-processing organ capable of absorbing enormous bodies of information that did not exist throughout our evolutionary history: chess gambits, stock prices, baseball records, organic chemical syntheses, and so on. Each of these bodies of information becomes the mental world of different individuals in our society. To assess the biophilia hypothesis, we shall have to identify and evaluate evidence that acquisition of knowledge about the natural world might have an innate basis lacking in the acquisition of knowledge about our worlds of human artifacts.

VIEWPOINT 5

"To continue blindly in this direction . . . is to flirt with the ultimate disaster—destruction of the earth as a fit living place for humans."

Humans Are Endangering Themselves

Robert M. McClung

In the following viewpoint, Robert M. McClung argues that humans may well be driving themselves toward extinction. Human-created problems such as pollution, overpopulation, and overconsumption are making the earth an increasingly uninhabitable place for all forms of life, McClung asserts. Unless human beings rapidly change their present course, he warns, humanity may soon be an endangered species. McClung is the author of *Lost Wild America: The Story of Our Extinct and Vanishing Wildlife*, from which this viewpoint is excerpted.

As you read, consider the following questions:

1. How did humans grow increasingly out of touch with nature, in McClung's opinion?
2. What harmful effects could result from the buildup of carbon dioxide in earth's atmosphere, according to the author?
3. In McClung's opinion, how have people misinterpreted the passage he quotes from Genesis?

Excerpted from *Lost Wild America: The Story of Our Extinct and Vanishing Wildlife*, Revised, Expanded, and Updated Edition, by Robert M. McClung (North Haven, CT: Linnet Books, 1993) by permission.

At this moment in history, *Homo sapiens* appears to be one of the most successful species the world has ever known. From humble beginnings human beings have risen to become the dominant form of life on earth, the one species that can alter and manipulate its environment to suit itself. Our superior brain and reasoning power made such a rise possible.

Human Development

In the beginning, this application of brainpower toward a different kind of life was very slow. For hundreds of thousands of years our distant ancestors lived as foraging and hunting animals. Just one of many different forms of life struggling for survival, they lived with and by the land. Every one of their senses was attuned, as the senses of the other animals still are, to all the rewards and dangers of their natural surroundings.

Very gradually they became herdsmen and tillers of the soil. They invented tools and fashioned weapons. They organized themselves in tribes, settled down in villages, and eventually built great cities and cultivated the arts. Over many thousands of years they evolved into what they are today: creatures increasingly out of touch with the natural world of their beginnings.

During the twentieth century this process of change has accelerated at a dizzying pace, as a result of the explosion of scientific knowledge and the application of that knowledge to medicine, industry, agriculture, and every other facet of modern life. "In the last few decades, mankind has been overcome by the most fateful change in its entire history," Barbara Ward declares in her book *Spaceship Earth*. "Modern science and technology have created so close a network of communications, transport, economic interdependence—and potential nuclear destruction—that planet earth, on its journey through infinity, has acquired the intimacy, the fellowship, and the vulnerability of a spaceship."

But most of us glorify the concept of the spaceship in quite a different way. Today everything seems possible, and we reach for the stars.

The Future

What do the swift tides of change augur for human life in the future? Not surprisingly, opinions vary. Technology advocates wax enthusiastic over their vision of the twenty-first century, declaring that applied science and technology will make all Earth a utopia, with the land transformed so that it can support many times the present population. They envision a brave new world of scientific triumphs: gleaming cities, all enclosed under domes, where the environment will be controlled and purified; people living in mile-high, connected towers, where no one will need to venture outside for weeks or months at a time. If they should, giant rocket-propelled transports will enable them to whisk from New York to London, Moscow, or Beijing within minutes. Food will be produced in chemical nutrients on vast indoor farms. There will be underseas farms too, where seaweed and plankton and other seafood can be continually grown and harvested. Disease will be conquered, and scientists will be able to manipulate human genes to produce "desirable" types of human beings at will.

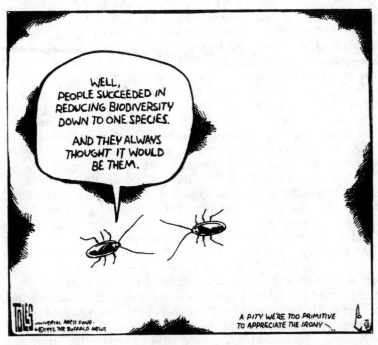

Toles. ©1992 *The Buffalo News*. Reprinted with permission of Universal Press Syndicate. All rights reserved.

All of these marvels are indeed possible—and that very fact profoundly disturbs a great many people. What will be the effect of such a machine-dominated civilization, not only upon our natural environment but upon our soul and spirit as well?

Already there are too many people in the world. Yet, along with the explosion of knowledge, the human population of the world keeps increasing. Every year there are at least ninety million more people on the planet Earth than there were the year before: ninety million more individuals that need more food, more homes, more products—and more technology to produce these things.

Little wonder that progress and prosperity in the advanced nations are generally equated with a constantly growing gross national product, and that in the frantic effort to meet the demands of an insatiable economy, we carry the raid on our national resources, the destruction of our natural environment, further and further every day. "With our spanking new toy, technology, we have already done more to disrupt natural things in our own lifetimes than were previously disrupted by all living things, including man, in all previous history," David Brower, now chairman of Earth Island Institute, asserted a quarter of a century ago in *Technology Review*. "Can we go on this way, worshipping growth, confusing it with progress?" How true those words ring today!

To continue blindly in this direction, ecologists warn, is to flirt with the ultimate disaster—destruction of the earth as a fit living place for humans and every other living thing. We have already advanced so far down this path of "more people, more technology" that no matter what programs may be launched to reverse the trend, they will not be enough to prevent a great deal of agony in the world brought about by famines, overcrowding, and the struggle to survive in a deteriorated environment. As Dr. Edward O. Wilson noted in a 1993 article in *The New York Times Magazine*, "On November 18, 1992, more than 1,500 senior scientists from 69 countries issued a 'Warning to Humanity' stating that overpopulation and environmental deterioration put the very future of life at risk."

Tipping the Ecological Balance

Heedless or unaware of such a possibility, most people continue on their accustomed way, demonstrating ever more forcefully that *Homo sapiens*, in relation to his environment, is the world's dirtiest and most destructive species. "A case could certainly be made out for the contention that modern man as a race has the death wish," Joseph Wood Krutch observed in a 1967 article. "Otherwise he would not be marching so resolutely toward literal extinction."

Consider:

The Sierra Club Legal Defense Fund notes that "In the United States alone, we dump 2.7 billion pounds of toxics into the air every year, discharge over 500 million pounds of toxic waste into our rivers, and bury 160 million tons of trash. Worldwide, we annihilate 74,000 acres of rain forest a day . . . 17,500 plant and animal species a year."

In a similar vein, Al Gore notes that every man, woman, and child in our country produces twice his or her weight in wastes—garbage, sewage, cans, paper, plastic, and many other throwaways—every *day*.

It is impossible to predict the long-range consequences of our fossil-fuel civilization, based on the vast consumption of coal and oil, which is adding carbon compounds to the atmosphere at a rate more rapid than plants on land and in the oceans can absorb them by photosynthesis. Just one big jetliner, it has been calculated, burns about six tons of petroleum hydrocarbons for every hour of flight and releases into the atmosphere about eight tons of water and twice that amount of carbon dioxide. Multiply this by thousands of planes, twenty-four hours in a day, 365 days in a year. Then add the carbon compounds released into the air from all other sources, and some idea of the extent of the problem can be gained.

By means of photosynthesis, plants utilize carbon dioxide to manufacture organic compounds, releasing oxygen as a byproduct. Animals, on the other hand, breathe oxygen and release carbon dioxide as a waste product of metabolism. The vital carbon-oxygen relationship has remained essentially in balance for some 400 million years of earth history. But now, through our use of fossil fuels, we are increasing the amount of carbon dioxide in the atmosphere, and at the same time, through bulldozer, ax, plow, and pollution, we are reducing the earth's potential for oxygen production. As a result, we are upsetting the earth's carbon-oxygen balance and altering the basic chemical, geological, and biological cycles on which all life depends.

Evidence of Destruction

Every year we see increasing evidence of the destruction we are inflicting on planet Earth from the millions of tons of toxic gases we are pumping into the atmosphere. In 1988 alone, we put 5.5 million tons of carbon into the atmosphere by burning fossil fuels. Scientists warn us that the buildup of carbon dioxide and other gases is causing a "greenhouse effect" and consequent global warming, since the opacity of the carbon dioxide inhibits the earth's surface heat from radiating back into space. Temperature increases of anywhere from 2 to 9 percent are predicted for the twenty-first century, with melting of much of the polar icecaps, rising sea levels, drastic climate changes over much of the earth, and other unpleasant and potentially disastrous possibilities.

Scientists also view with alarm the "ozone holes" that are appearing over both the Antarctic and arctic regions. The thinning of the protective ozone layer in the atmosphere allows increased deadly ultraviolet radiation to get through to the earth. It is caused for the most part by the emission of chlorofluorocarbons (CFCs) into the atmosphere, where they break down into ozone-destroying chemicals. In 1989 the United States alone produced 700 million pounds of CFCs, which are used in refrigerants, air conditioners, and aerosol sprays. Another environmentally destructive force is acid rain, caused by sulfate and nitrate emissions into the atmosphere. In the air, the nitrogen and sulfur compounds combine with oxygen to form acids, which are then removed from the air by rain or snow which falls upon our forests and lakes. Acid rain kills or stunts the trees and renders thousands of lakes in the United States and Canada unfit for fish, amphibians, and other aquatic life.

Assessing the continuing deterioration of the environment, the Worldwatch Institute, a Washington, D.C.-based environmental research organization, declared in its annual report, *State of the World, 1993*, that "nothing short of sharp changes in government policies and people's attitudes will rescue the earth's ailing ecosystems from destruction.". . .

Diminishing Human Chances

As our activities continue to wipe out plants and animals, we may be unknowingly diminishing human chances of survival, through loss of critical medicines, food supplies, and the general balance of nature. For want of a dam, a ranch, a lumber supply, or yet another housing development, another species may eventually be lost. Our own.

Noel Grove, *American Forests*, November/December 1992.

All of the environmental pollution that threatens to upset the earth's ecological balance follows in the wake of a civilization based on the ideal of making more and more *things* for more and more *people*. In the process, our sense of values has been distorted, and we seem to have lost our feeling of kinship with the natural world that has sustained us for so long. Vice President Al Gore sadly observes that the enshrined political ethic of the age seems to be, "Get it while you can; forget about the future." In 1967, sociologist Richard L. Means, in an article in *Saturday Review*, pointed out that

> justification of a technological arrogance toward nature on the basis of dividends and profits is not just bad economics—it is basically an immoral act. And our contemporary moral crisis, then, goes much deeper than questions of political power and law, or urban riots and slums. It may, at least in part, reflect American society's almost utter disregard for the value of nature.

Part of America's arrogant attitude toward nature is embedded in the mystique of the pioneer spirit: man against nature, rugged individualism, every man for himself. Part of it, perhaps, goes much farther back, to the beginnings of Western culture and ethics. Albert Schweitzer once observed that "the great fault of all ethics hitherto has been that they believed themselves to have to deal only with the relations of man to man."

The root of the crisis, perhaps, lies in the widely held but misleading premise that all of nature exists only to serve our use. For many centuries there has been a general misinterpretation of the passage in the Bible's Book of Genesis in which God told man: "Have dominion over the fish of the sea, and the fowl of the air, and over every living thing that moveth upon the earth." The understanding of this passage has usually emphasized man's dominance, his right to use and deal with all other animals as he wishes. A more enlightened interpretation points out our moral responsibility for safeguarding wildlife and all the rest of Earth's resources, and using them wisely.

Human beings could, of course, survive in a world without whooping cranes or California condors, grizzly bears or blue whales. Our very existence does not depend upon these and some others of our fellow inhabitants of the earth. But if we should ever become indifferent to the question of whether these animals live or not, if we didn't recognize the worth of saving them, then we would no longer be human in terms of spirit and moral essence.

The question of wildlife preservation also could be answered on the level of simple self-interest: the dwindling fortunes of our wildlife serve as warning signals of mankind's possible fate too. If we cannot save endangered wildlife by preserving some of its essential environment and conditions for its survival—then the human race, in the long run, will go under too.

VIEWPOINT 6

"It's time to give up the notion of human beings as intruders, tramplers, and destroyers."

Humans Are Not Endangering Themselves

Thomas Palmer

Human beings are not placing their own species in danger, Thomas Palmer argues in the following viewpoint. Palmer asserts that humans have erroneously come to see themselves as being outside of nature and even as a threat to it. Rather, Palmer suggests, the changes that humans have made to their natural environment contribute to the normal cycle of species evolution and extinction. Nature, he says, not humanity, will ultimately determine the progress of the human species. Palmer is the author of several books, including *Landscape with Reptile: Rattlesnakes in an Urban World*, from which this viewpoint was adapted.

As you read, consider the following questions:

1. According to the author, how did the Puritans affect their environment?
2. Why don't humans fit neatly into ecological models, in Palmer's opinion?
3. In what respect does Palmer compare the earth to a fertile egg?

Excerpted from *Landscape with Reptile*, by Thomas Palmer, as published in *The Atlantic Monthly* (January 1992). Copyright ©1992 by Thomas Palmer. Reprinted by permission of Ticknor & Fields/Houghton Mifflin Co. All rights reserved.

An argument, a human argument, maintains that we ought to be concerned about the disappearance of individual animal species. If it could be directed at the objects of its solicitude, it would go approximately as follows: "You lesser beasts had better watch your step—*we'll* decide when you can leave." It recognizes that once chromosome patterns combine at the species level, they become unique and irreplaceable—one cannot make a rattlesnake, for instance, out of anything but more rattlesnakes. It looks at the speed at which such patterns are disappearing and shudders to think how empty our grandchildren's world might become, patternwise.

Since the early 1970s, this argument has conquered much of the world; it may soon become part of the thinking of nearly every school child.

Perhaps because we ourselves are a species, we regard the species level as that at which deaths become truly irreversible. Populations, for instance, can and do fade in and out; when a species dies, however, we call it extinct and retire its name forever, being reasonably certain that it will not reappear in its old form.

Students of evolution have shown that species death, or extinction, is going on all the time, and that it is an essential feature of life history. Species are adapted to their environments; as environments change, some species find themselves in the position of islanders whose islands are washing away, and they go under. Similarly, new islands (or environments) are appearing all the time, and they almost invariably produce new species.

Serious Questions

What alarms so many life historians is not that extinctions are occurring but that they appear to be occurring at a greater rate than they have at all but a few times in the past, raising the specter of the sort of wholesale die-offs that ended the reign of the dinosaurs. Do we want, they ask, to exile most of our neighbors to posterity? Exactly how much of our planet's resources do we mean to funnel into people-making? Such questions are serious; they involve choosing among futures, and some of these futures are already with us, in the form of collapsing international fisheries, rich grasslands gnawed and trampled into deserts, forests skeletonized by windborne acids, and so forth. Thus high rates of extinction are seen as a symptom of major problems in the way our species operates—problems that may, if we're not careful, be solved for us. A new word has been coined to define the value most threatened by these overheated rates: "biodiversity." As species disappear, biodiversity declines, and our planet's not-quite-limitless fund of native complexities—so some argue—declines with it.

The process described above is indeed occurring. Human beings tend to change environments; when they do, species vanish. The Puritans, for example, though famous for their efforts to discipline sexuality, imposed upon Massachusetts an orgy of ecological licentiousness: they introduced dozens of microbes, weeds, and pests foreign to the region, some of which played havoc with the natives. Human beings tend to travel everywhere, and to bring their cats, rats, and fleas with them, so that hardly any environment is truly isolated today, and creatures that evolved in isolated environments have paid a high price. Of the 171 species and subspecies of birds that have become extinct in the past 300 years, for example, 155 were island forms.

Beyond Human Control

Nature is already saved and, moreover, largely out of our hands. If once we thought all organisms were for our benefit, and later we thought we could with bombs kill off all life on the planet, it is once again a mark of our hubris to think that we may now save the biological world. . . .

In fact, we cannot stop evolution. We can, and probably should, try to stop certain global human activities among which may be counted overuse of plastics, rain forest destruction, and soil erosion. But to think that by doing so or not we are either going to kill off life on earth or save it is a form of unscientific self-aggrandizement.

Dorion Sagan and Lynn Margulis, *The Biophilia Hypothesis*, edited by Stephen R. Kellert and Edward O. Wilson, 1993.

Since extinction is a particularly final and comprehensive form of death, species preservation and its corollary, habitat protection, are now seen as the most important means available to stem the erosion of biodiversity. So far, so good—but I wonder if these ideas, which emphasize diversity at the species level, fail to give an adequate picture of recent biological history. If, for instance, biodiversity is regarded as the chief measure of a landscape's richness, then the American continents reached their peak of splendor on the day after the first Siberian spearmen arrived, and have been deteriorating ever since. More recent developments—such as the domestication of maize, the rise of civilizations in Mexico and Peru, and the passage of the U.S. Bill of Rights—are neutral at best, and are essentially invisible, since they are the work of a single species, a species no more or less weighty than any other, and already present at the start of the interval. But what kind of yardstick measures a handful of skin-clad hunters against Chicago, Los Angeles, and Caracas, and finds one group no more "diverse" than the other?

A considerable amount of pessimism is built into this species-based notion of diversity. Nearly all change on such a scale is change for the worse—especially human-mediated change. Change involves stress, and stress causes extinctions; each extinction is another pock in the skin of an edenic original. This original is frozen in time; more often than not, it is defined as the blissful instant just prior to the arrival of the first human being. In fact, the only way to re-create this instant, and restore biodiversity to its greatest possible richness, would be to arrange for every human being on earth to drop dead tomorrow.

This is not to say that cities are better than coral reefs, or that binary codes are an improvement on genetic ones, but only that "biodiversity" cannot adequately account for the phenomenon of *Homo sapiens*.

Maybe it's time to give up the notion of human beings as intruders, tramplers, and destroyers. We are all of these, there's no doubt about it, but they are not all we are. And yet the same mind-set that interprets human history as little more than a string of increasingly lurid ecological crimes also insists that our species represents the last, best hope of "saving" the planet. Is it any wonder that the future looks bleak? . . .

Humans' Place

Ecologists, biologists, and environmentalists have had fits trying to introduce our species into their models of the natural world. These models are based on the idea of balance, or equilibrium, wherein each variety of plant or animal plays a limited, genetically prescribed role in the cycling of materials and energy. The roles are not absolutely fixed—natural selection, by sorting and resorting chromosomes, can adapt lines of descent to new ones—but change, by and large, is assumed to be gradual, and millions of years can pass without any notable restructuring of communities.

Human beings cannot be worked into such models. One cannot look at human beings and predict what they will eat, or where they will live, or how many of their children a given landscape will support. If they inhabit a forest, they may burn it down and raise vegetables, or flood it and plant rice, or sell it to a pulp-and-paper manufacturer. They may think of anything; the life their parents led is not a reliable blueprint, but merely a box with a thousand exits. Moralists in search of instructive contrasts will sometimes idealize primitive societies, claiming that they deliberately live "in balance" with their environments, but these examples don't stand up to scrutiny. The Massachuset Indians, for instance, though sometimes presented as sterling conservationists, were the descendants of aboriginal American hunters who appear to have pursued a whole constellation of Ice Age mammals to extinction (including several species of horses). When, in historical times, they were offered metal fish-hooks, knives, and firearms, they didn't say, "Thanks, but we prefer rock-chipping."

The revelation that we are not like other creatures in certain crucial respects is an ancient one, and may be nearly as old as humanity; it probably contributed to the idea, central to several major religions, that we inhabit a sort of permanent exile. Until recently, however, we could still imagine ourselves encompassed by, if not entirely contained in, landscapes dominated by nonhuman forces—weather, infectious illness, growing seasons, light and darkness, and so forth. This is no longer so; today most human beings live in artificial wildernesses called cities, and don't raise the food they eat, or know where the water they drink fell as rain. A sort of vertigo has set in—a feeling that a rhythm has been upset, and that soon nothing will be left of the worlds that made us. This feeling is substantiated by population curves, ocean pollution, chemical changes in the earth's atmosphere, vanishing wildlife, mountains of garbage, and numerous other signs that anyone can read. The nineteenth-century conservation movement, which sought to preserve landscapes for largely aesthetic reasons, has become absorbed in the twentieth-century environmental movement, which insists that more is at stake than postcard views. We are, it argues, near to exceeding the carrying capacity of our planet's natural systems, systems whose importance to us will become very obvious when they begin to wobble and fail.

These are not empty warnings. Human communities can and occasionally do self-destruct by overstraining their resource bases. Historical examples include the Easter Islanders, the lowland Maya, and some of the classical-era city-dwellers of the Middle East and North Africa. But if we set aside the equilibrium-based models of the ecologists, and do not limit ourselves to species-bound notions of diversity—in other words, if we seek to include human beings in the landscape of nature, rather than make them outcasts—what sort of picture do we get of the phenomenon of life? . . .

The Uniqueness of Life

Life, for the biologists, is an uphill or retrograde process—it adds order and complexity to environments whose overall tendency is toward diffusion and disorder. It captures energies released by decay and exploits them for growth and rebirth. It is startlingly anomalous in this respect: so far as we know, it occurs nowhere but on the surface of this planet, and even here its appearance seems to have been a one-time-only event; though many lifelike substances have been produced inside sterile glassware, none has ever quickened into veritable beasthood.

The evidence suggests that life continued to fructify and elaborate itself for several billion years after its appearance. The milestones along the way—the nucleated cell, photosynthesis, sexual reproduction, multicellularity, the internal skeleton, the invasion of the land and sky, and so forth—are usually interpreted as advances, because they added additional layers of complexity, interconnection, and ordered interaction to existing systems. This drama did not proceed without crises—photosynthesis, for instance, probably wiped out entire ecosystems by loading the atmosphere with a deadly poison, free oxygen—but life as a whole laughed at such insults, and continued on its protean way.

If we believe that all life—in contrast to rocks and gases—shares a certain quality of sensitivity, or self-awareness, then *Homo sapiens* was an astonishing and wholly unpredictable leap forward in this respect, because human beings manifested an idea of personhood never before achieved. . . .

Human Qualities

Consciousness. Mind. Insight. Here are qualities that, if not exclusively human, seem appallingly rudimentary elsewhere. Primitive peoples distributed them throughout their worlds; we moderns hold to stricter standards of evidence. Does a cloud yearn, for instance, to drop rain? Is a seed eager to sprout?

The irruption of thoughtfulness that our species represents is not inexplicable in Darwinian terms. Once our apelike and erect ancestors began using weapons, hunting large animals, and sharing the spoils, the ability to develop plans and communicate them acquired considerable survival value, and was genetically enhanced. This ability, and the tripling in brain weight that accompanied it, turned out to be one of the most revolutionary experiments in the history of gene-sorting. . . .

If life . . . is a paradoxical chemical process by which order arises from disorder, and a movement toward uniformity produces more-complex local conditions, then human enterprise, though full of disasters for other species, is clearly not outside the main line of development. Equatorial rain forests, for instance, are probably the most diverse and multifaceted communities of species on earth. But are they more densely stuffed with highly refined codes and labels than, say, the Library of Congress? Long ago certain moths learned to communicate over as much as two miles of thick woods by releasing subtle chemicals that prospective mates could detect at levels measured in parts per million; today a currency broker in Tokyo can pick up a phone and hear accurate copies of sounds vocalized a split second earlier by a counterpart on the other side of the world. Which system of signals is more sensitive and flexible?

The Crown of Creation

I am concerned, as is obvious with an image—the image of our species as a vast, featureless mob of yahoos mindlessly trampling this planet's most ancient and delicate harmonies. This image, which is on its way to becoming an article of faith, is not a completely inaccurate description of present conditions in some parts of the world, but it portrays the human presence as a sort of monolithic disaster, when in fact *Homo sapiens* is the crown of creation, if by creation we mean the explosion of earthly vitality and particularity long ago ignited by a weak solution of amino acids mixing in sunlit waters. Change—dramatic, wholesale change—is one of the most reliable constants of this story. To say that the changes we have brought, and will continue to bring, are somehow alien to the world, and are within a half inch of making its "natural" continuance impossible, displays some contempt, I think, for the forces at work, along with a large dose of inverted pride. Who are we, for instance, to say what's possible and what isn't? Have we already glimpsed the end? Where exactly did things go awry? It's useful to remember that just yesterday our main concern was finding something to eat.

I prefer to suppose that we will be here awhile, and that such abilities as we have, though unprecedented in certain respects, are not regrettable. The human mind, for instance, could never have set itself the task of preserving rare species if earlier minds had not learned how to distinguish light from darkness, or coordinate limbs, or identify mates. Now that we think we know something about our immediate neighborhood, we are beginning to realize what a rare quality life is, and if we think of its multibillion-year history on earth as a sort of gradual awakening of matter, we must conclude that the dawning of human consciousness represents one of the most extraordinary sunrises on record. Is it any wonder, then, that the world is changing?

Perhaps because we have become so expert at interrogating our surroundings, we tremble a little at our own shadows. God, for instance, has become almost a fugitive. We have disassembled the atom; we have paced off the galaxies; He doesn't figure in our equations.

Maybe it would be useful at this point to compare our common birthplace to a fertile hen's egg. Nearly everyone has seen the delicate tracery of blood vessels that begins to spread across the yolk of such an egg within a few hours of laying. Before long a tiny pump starts to twitch rhythmically, and it drives a bright scarlet fluid through these vessels. The egg doesn't know that it is on its way to becoming a chicken. Chickens, for the egg, lie somewhere on the far side of the beginning of time. And yet the egg couldn't be better equipped to make a chicken out of itself.

Doomsayers

I would argue that our planet, like the egg, is on a mission of sorts. We don't know what that mission is any more than the nascent nerve cells in the egg know why they are forming a network. All we know is that things are changing rapidly and dramatically.

Today many believe that these changes are often for the worse, and represent a fever or virus from which the body of life will emerge crippled and scarred. We look back with longing on a time, only a moment ago, when the human presence barely dimpled the landscape—when the yolk, so to speak, was at its creamiest, and no angry little eye-spots signaled an intent to devour everything.

I'm not persuaded by this picture—I think it arises from a mistaken belief that the outlines of earthly perfection are already evident. It has inspired a small army of doomsayers—if we burn the forests of the Amazon, we are told, our planet's lungs will give out, and we will slowly asphyxiate. Surely we have better, more practical reasons for not burning them than to stave off universal catastrophe. I can easily imagine similar arguments that would have required the interior of North America to remain empty of cities—and yet I don't think this continent is a poorer place now than it was 20,000 years ago. The more convinced we are that our species is a plague, the more we are obliged to yearn for disasters.

Students of historical psychology have noticed that the end of the world is always at hand. For the Puritan preachers it was to take the form of divine wrath, and they warned that the Wampanoag war [1675–76 war between New England colonists and the Wampanoag tribe; also called King Philip's War] was only a foretaste. The Yankees saw it coming in the flood of nineteenth-century immigrants, who meant to drown true Americanism. Today we are more likely to glimpse it in canned aerosols, poisoned winds, and melting ice caps.

Curiously enough, the end of the world always *is* at hand—the world dies and is reborn on a daily basis. A fertile hen's egg is never today what it was yesterday, or will be tomorrow. Few would deny that the effort to preserve and protect as many as possible of the millions of species now existing represents a fresh and heartening expansion of human ambitions. But to suppose that earthly diversity is past its prime, and that a strenuous program of self-effacement is the best contribution our species has left to offer, is neither good biology nor good history.

Periodical Bibliography

The following articles have been selected to supplement the diverse views presented in this chapter.

Warren D. Allmon	"What Are We Doing to Evolution?" *The World & I*, December 1990. Available from 2800 New York Ave. NE, Washington, DC 20002.
Betsy Carpenter	"Living with Nature," *U.S. News and World Report*, November 30, 1992.
Alan Thein Durning	"Return of the Native," *World Monitor*, March 1993. Available from One Norway St., Boston, MA 02115.
Arturo Gomez-Pompa and Andrea Kraus	"Taming the Wilderness Myth," *Bioscience*, April 1992.
Noel Grove	"The Species You Save May Be Your Own," *American Forests*, November/December 1992. Available from 1516 P St. NW, Washington, DC 20005.
William Hazeltine	"A Petition to Protect Endangered *Homo Sapiens*," *21st Century Science and Technology*, Winter 1994–95. Available from PO Box 16285, Washington, DC 20041.
Joe Kane	"Moi Goes to Washington," *The New Yorker*, May 2, 1994.
Matt Moffett	"Kayapo Indians Lose Their 'Green' Image," *The Wall Street Journal*, December 29, 1994.
Paul C. Pritchard	"The Evolution Time Bomb," *National Parks*, July/August 1993.
Propaganda Review	"Against Nature: The Ideology of Ecocide," Spring 1994.
UN Chronicle	"The Development Dilemma: Sustaining Resources, Improving Livelihoods," June 1993.
Robert Weissman	"Disappearing Trees, Disappearing Culture," *Multinational Monitor*, April 1994.
Edward O. Wilson	"Is Humanity Suicidal?" *The New York Times Magazine*, May 30, 1993.
Davi Kopenawa Yanomami	"Yanomami in Peril," *Multinational Monitor*, September 1992.

Glossary

adaptive radiation   The evolution of a single **species** into a number of adaptively specialized species.

anthropogenic   Relating to or resulting from the influence of humans on nature.

biodiversity   Biological diversity in an environment as indicated by the variety of living organisms present.

breed   A group of domesticated animals related by descent from common ancestors; analogous to **subspecies**.

critical habitat   The minimum amount of land required to halt a **species'** population decline.

ecosystem   An ecological unit of nature that consists of a specific physical environment and the organisms living in that environment.

endemic   A **species** peculiar to its native environment and not found elsewhere.

exotic   A **species** that is not native to the environment where it is found.

faculative   Exhibiting certain characteristics under some environmental conditions but not under others.

fauna   All the animals found in a specific environment.

genetic erosion   Reduction or loss of genetic diversity over time.

genus (plural: **genera**)   A group of closely related **species**.

germplasm   The hereditary material of germ cells.

habitat   An environment of a particular kind; specifically, the area in which a plant or animal normally lives.

hybrid   The offspring of two animals or plants that are genetically dissimilar, especially of parents that belong to different **species** or **genera**.

indigenous   Native to a particular region or environment; in reference to humans, the term often implies native people who adhere to their traditional customs.

montane   Relating to or living in a mountain environment that consists of cool upland slopes below the timberline.

riparian   Relating to or living on the bank of a natural waterway.

species   A biological category comprising closely related and similar organisms.

subspecies   A subdivision of a **species**; a population living in a particular region that is genetically different from other populations of the same species.

taxon (plural: **taxa**)   Each separate level in the classification of organisms, e.g., **species** and **genus**.

taxonomy   Specifically, the classification of animals and plants according to their natural relationships; sometimes refers to all aspects of the origins and contents of **biodiversity**.

Vavilovian center   A region that contains plants in both wild and cultivated states; a center of unusual genetic diversity in a given plant **species**.

For Further Discussion

Chapter 1

1. Julian L. Simon and Aaron Wildavsky claim that Edward O. Wilson's evidence for rapid species extinction is based only on "anecdotal reports" that are of "little or no value." Do you find their criticism of Wilson's citations to be valid? Why or why not?

2. John C. Ryan and Shawn Carlson both suggest ways to benefit local economies while preserving rain forest areas and maintaining acceptable levels of biodiversity. In your opinion, which author presents better methods for balancing human needs and conserving the rain forest? Explain.

3. The viewpoints of Lisa Drew and Al Gore are largely concerned with the threat of extinction to domesticated animals and agricultural plants. Which do you believe is more urgent—saving endangered wildlife or protecting endangered domestic breeds? On what do you base your answer?

Chapter 2

1. Robert E. Gordon argues that the Endangered Species Act has been ineffective because few species have been removed from the endangered species list. What arguments does Michael J. Bean give to support the efficiency of the Endangered Species Act?

2. Considering the viewpoints of Colin Tudge and Fiona Sunquist, would you recommend wildlife reintroduction programs as a viable method for preserving some endangered species? Would you recommend it for all endangered species?

3. Roger L. DiSilvestro contends that humans should not "play God" by deciding to withhold aid from endangered species that are otherwise likely to become extinct. Suzanne Winckler asserts that people must allow some species to become extinct so that the larger majority can be saved. What facts do the authors give to support their arguments? Which parts of each author's argument rely on emotional appeals?

Chapter 3

1. Randy Fitzgerald uses unemployment statistics to bolster his argument that saving the northern spotted owl's habitat has cost thousands of lumber-industry jobs. Alexander Cockburn and Timothy Egan also quote statistics to support their contention that protecting the northern spotted owl has not caused substantial unemployment of loggers. Whose use of statistics do you find more convincing? Why?

2. Alston Chase and the editors of *Human Events* assert that reintro-

duction of wolves will harm ranchers' livelihoods. In her viewpoint, Renée Askins maintains that wolves are unlikely to seriously menace humans or their livestock. Do you believe that Askins successfully counters the arguments of Chase and *Human Events*? Why or why not?

3. William F. Jasper cites examples of private citizens who have faced stiff legal penalties. Douglas A. Thompson and Thomas G. Yocom contend that most opponents of wetland regulations are not small landowners but large companies such as Exxon. In your opinion, should wetland regulations apply equally to private individuals and major corporations, or should certain allowances be made to either? Explain your reasoning.

Chapter 4

1. The Norwegian author Mari Skåre addresses a predominantly American audience on the issue of lifting the commercial whaling ban. What differences between Norwegian and U.S. cultures does Skåre emphasize? Does knowing the attitude of the "average Norwegian" toward whales and whaling affect your opinion about the whaling ban? Explain.

2. What solutions do Andre Carothers and Richard C. Morais suggest to Africa's endangered wildlife problem? Evaluate the possible effectiveness and drawbacks of these solutions. Propose some solutions of your own.

3. Based on your consideration of the viewpoints of Diane Jukofsky and Stephen Corry, would you choose to buy products that are advertised as containing ingredients harvested from the rain forest? Why or why not?

Chapter 5

1. To support his argument that the indigenous peoples of the Amazon are endangered, Art Davidson includes extensive quotes from native leaders who are working to protect their way of life. Do you believe the addition of these leaders' opinions strengthens or weakens Davidson's viewpoint? Explain.

2. In his theory of biophilia, Edward O. Wilson proposes that humans across the world tend to share certain feelings toward nature. On the other hand, Jared Diamond presents examples of Stone Age peoples who do not necessarily exhibit these traits. In your opinion, are Diamond's findings sufficient to disprove Wilson's theory of biophilia? Why or why not?

3. Robert M. McClung cites statistics on environmental destruction to support his contention that the human species is endangering itself. Thomas Palmer uses examples of human ingenuity and resourcefulness to bolster his assertion that human civilization is a positive part of life on earth. Which of these arguments do you find more effective? Explain.

Organizations to Contact

The editors have compiled the following list of organizations concerned with the issues debated in this book. The descriptions are derived from materials provided by the organizations themselves. All have publications or information available for interested readers. The list was compiled on the date of publication of the present volume; names, addresses, and phone numbers may change. Be aware that many organizations take several weeks or longer to respond to inquiries, so allow as much time as possible.

American Forest and Paper Association (AFPA)
1250 Connecticut Ave. NW, 2nd Fl.
Washington, DC 20036
(202) 463-2455

AFPA is a national trade association of the forest, pulp, paper, paperboard, and wood products industry. The association publishes materials on timber supply and forest management as well as the *International Trade Report*, a monthly newsletter that features articles on current issues affecting forest products, industry, and international trade.

American Livestock Breeds Conservancy (ALBC)
Box 477
Pittsboro, NC 27312
(919) 542-5704

ALBC works to prevent the extinction of rare breeds of American livestock. The conservancy believes conservation is necessary to protect the genetic range and survival ability of these species. ALBC provides general information about the importance of saving rare breeds as well as specific guidelines for individuals interested in raising rare breeds.

American Zoo and Aquarium Association (AZA)
7970 Old Georgetown Rd., Suite D
Bethesda, MD 20814
(301) 907-7777
fax: (301) 907-2980

AZA represents over 160 zoos and aquariums in North America. The association provides information on captive breeding of endangered species, conservation education, natural history, and wildlife legislation. AZA publications include the *Species Survival Plans* and the *Annual Report on Conservation and Science*. Both publications are available from the Office of Membership Services, Oglebay Park, Wheeling, WV 26003-1698.

Canadian Forestry Association (CFA)
185 Somerset St. W, Suite 203
Ottawa, ON K2P 0J2
CANADA
(613) 232-1815
fax: (613) 232-4210

CFA works for improved forest management that would satisfy the economic, social, and environmental demands on Canadian forests. The association explores conflicting perspectives on forestry-related topics in its biannual *Forest Forum*.

Center for Plant Conservation (CPC)
Missouri Botanical Garden
PO Box 299
St. Louis, MO 63166
(314) 577-9450

CPC is a network of twenty-five botanical gardens and arboreta concerned with plant conservation. The center gathers and disseminates information on endangered plants indigenous to the United States and conserves seeds and cuttings of rare plants to preserve their genetic patterns. CPC publications include the biannual newsletter *Plant Conservation* and the annual *Plant Conservation Directory*.

Cultural Survival (CS)
46 Brattle St.
Cambridge, MA 02138
(617) 441-5400
fax: (617) 441-5417

Cultural Survival strives to help the indigenous people of undeveloped regions to survive, both physically and culturally, the rapid changes brought on by contact with industrial societies. It serves as a center for research and documentation on the problems facing indigenous peoples and threatened societies. In addition to its numerous books and reports, the organization publishes the magazine *Cultural Survival Quarterly*.

Endangered Species Coalition (ESC)
666 Pennsylvania Ave. SE
Washington, DC 20003
(202) 547-9009

The coalition is composed of conservation, professional, and animal welfare groups that work to extend the Endangered Species Act and to ensure its enforcement. ESC encourages public activism through grassroots organizations, direct lobbying, and letter-writing and telephone campaigns. Its publications include the book *The Endangered Species Act: A Commitment Worth Keeping* and articles, fact sheets, position papers, and bill summaries regarding the Endangered Species Act.

Foundation for Research on Economics and the Environment (FREE)
502 S. 19th Ave.
Bozeman, MT 59715
(406) 585-1776

FREE is a research and education foundation committed to freedom, environmental quality, and economic progress. The foundation works to reform environmental policy by using the principles of private property rights, the free market, and the rule of law. FREE publishes the quarterly newsletter *FREE Perspectives on Economics and the Environment* and produces a biweekly syndicated op-ed column.

International Society of Tropical Foresters (ISTF)
5400 Grosvenor Ln.
Bethesda, MD 20814
(310) 897-8720
fax: (301) 897-3690

ISTF is a nonprofit international organization that strives to develop and promote ecologically sound methods of managing and harvesting tropical forests. The society provides information and technical knowledge about the effects of deforestation on agriculture, forestry, and industry. It publishes the quarterly newsletter *ISTF News*.

National Wildlife Federation (NWF)
1400 16th St. NW
Washington, DC 20036-2266
(202) 797-6800

Canadian Wildlife Federation (CWF)
2470 Queensview Dr.
Ottawa, ON K2B 1A2
CANADA
(800) 563-9453

These closely affiliated organizations encourage the conservation of natural resources in North America and throughout the world. Through summits, educational materials, and research, the NWF and CWF promote conservation and appreciation of wildlife. Publications include the annual *Conservation Directory*, the semiannual *Conservation Exchange*, and the bimonthly magazines *National Wildlife* and *International Wildlife*.

PERC
502 S. 19th Ave., Suite 211
Bozeman, MT 59715
(406) 587-9591
fax: (406) 586-7555

PERC is a research center that provides solutions to environmental problems based on free market principles and the importance of private property rights. PERC publications include the quarterly newsletter *PERC Report* and papers in the *PERC Policy Series* dealing with environmental issues.

Rainforest Action Network (RAN)
450 Sansome, Suite 700
San Francisco, CA 94111
(415) 398-4404

RAN works to preserve the world's rain forests and protect the rights of native forest-dwelling peoples. The network sponsors letter-writing campaigns, boycotts, and demonstrations in response to environmental concerns. It publishes miscellaneous fact sheets, the monthly *Action Alert* bulletin, and the quarterly *World Rainforest Report*.

South and Meso-American Indian Information Center (SAIIC)
PO Box 28703
Oakland, CA 94604
(510) 834-4263
fax: (510) 834-4264

SAIIC disseminates information about problems facing indigenous peoples in South and Central America. These problems include the increasing control of agriculture by large corporations and the damage done to the Amazonian rain forest by developers and others. SAIIC publishes the quarterly *Abya Yala News: Journal of the South and Meso American Indian Rights Center* and the book *Daughters of Abya Yala*.

United Nations Environment Programme (UNEP)
UNDC Two Building, Rm. 0803
Two United Nations Plaza
New York, NY 10017
(212) 963-8138
fax: (212) 963-7341

UNEP studies ecosystems, encourages environmental management and planning, and helps developing countries deal with their environmental problems. Its publications include environmental briefs, the bimonthly magazine *Our Planet*, and numerous books available through its publications catalogue.

U.S. Fish and Wildlife Service
Office of Public Affairs
1849 C St. NW
Washington, DC 20240
(202) 208-5634

The U.S. Fish and Wildlife Service is a network of regional offices, national wildlife refuges, research and development centers, national fish hatcheries, and wildlife law enforcement agents. The service's primary goal is to conserve, protect, and enhance fish and wildlife and their habitats. It publishes an endangered species list as well as fact sheets, pamphlets, and information on the Endangered Species Act.

World Wildlife Fund (WWF)
1250 25th St. NW
Washington, DC 20037
(202) 293-4800

World Wildlife Fund works to save endangered species, to conduct wildlife research, and to improve the natural environment. It publishes an endangered species list, the bimonthly newsletter *Focus*, and a variety of books on the environment.

Bibliography of Books

Lawrence Alderson and Robert Dowling	*Rare Breeds*. London: L. King, 1994.
Ron Arnold	*Trashing the Economy: How Runaway Environmentalism Is Wrecking America.* Bellevue, WA: Free Enterprise Press, 1994.
Ronald Bailey, ed.	*The True State of the Planet.* New York: Free Press, 1995.
Rocky Barker	*Saving All the Parts: Reconciling Economics and the Endangered Species Act.* Washington: Island Press, 1993.
Joseph L. Bast, Peter J. Hill, and Richard C. Rue	*Eco-Sanity: A Common-Sense Guide to Environmentalism.* Lanham, MD: Madison Books, 1994.
Raymond Bonner	*At the Hand of Man: Peril and Hope for Africa's Wildlife.* New York: Knopf, 1993.
Art Davidson	*Endangered Peoples.* San Francisco: Sierra Club Books, 1993.
Roger L. DiSilvestro	*Reclaiming the Last Wild Places: A New Agenda for Biodiversity.* New York: Wiley, 1993.
Iain and Oria Douglas-Hamilton	*Battle for the Elephants.* New York: Viking, 1992.
Andrea L. Gaski and Kurt A. Johnson	*Prescription for Extinction: Endangered Species and Patented Oriental Medicines in Trade.* Washington: World Wildlife Fund, 1994.
Brian Groombridge, ed.	*Global Biodiversity: Status of the Earth's Living Resources.* New York: Chapman & Hall, 1992.
Ginette Hemley, ed.	*International Wildlife Trade: A CITES Sourcebook.* Washington: Island Press, 1994.
Monte Hummel and Sherry Pettigrew with John Murray	*Wild Hunters: Predators in Peril.* Niwot, CO: Roberts Rinehart, 1992.
Les Kaufman and Kenneth Mallory	*The Last Extinction.* Cambridge, MA: MIT Press, 1993.
Wallace Kaufman	*No Turning Back: Dismantling the Fantasies of Environmental Thinking.* New York: BasicBooks, 1994.
Elizabeth Kemf	*The Law of the Mother: Protecting Indigenous Peoples in Protected Areas.* San Francisco: Sierra Club Books, 1993.
Kathryn A. Kohn, ed.	*Balancing on the Brink of Extinction: The Endangered Species Act and Lessons for the Future.* Washington: Island Press, 1991.

Thomas Lambert and Robert J. Smith	*The Endangered Species Act: Time for a Change.* St. Louis: Center for the Study of American Business, 1994.
Richard Littell	*Endangered and Other Protected Species: Federal Law and Regulation.* Washington: Bureau of National Affairs, 1992.
Charles E. Little	*The Dying of the Trees: The Pandemic in America's Forests.* New York: Viking, 1995.
Charles C. Mann and Mark L. Plummer	*Noah's Choice: The Future of Endangered Species.* New York: Knopf, 1995.
Peter H. Marshall	*Nature's Web: Rethinking Our Place on Earth.* New York: Paragon House, 1994.
Kenton Miller and Laura Tangley	*Trees of Life: Saving Tropical Forests and Their Biological Wealth.* Boston: Beacon Press, 1991.
Bryan G. Norton et al.	*Ethics on the Ark: Zoos, Animal Welfare, and Wildlife Conservation.* Herndon, VA: Smithsonian Institution Press, 1995.
Margery L. Oldfield and Janis B. Alcorn, eds.	*Biodiversity: Culture, Conservation, and Ecodevelopment.* Boulder, CO: Westview Press, 1991.
Catherine Paladino	*Our Vanishing Farm Animals: Saving America's Rare Breeds.* Boston: Joy Street Books, 1991.
Kathryn Phillips	*Tracking the Vanishing Frogs.* New York: St. Martin's Press, 1994.
Charles T. Rubin	*The Green Crusade: Rethinking the Roots of Environmentalism.* New York: Free Press, 1994.
George B. Schaller	*The Last Panda.* Chicago: University of Chicago Press, 1993.
Mark L. Shaffer	*Beyond the Endangered Species Act: Conservation in the 21st Century.* Washington: Wilderness Society, 1992.
Boyce Thorne-Miller and John G. Catena	*The Living Ocean: Understanding and Protecting Marine Biodiversity.* Washington: Island Press, 1991.
Richard Tobin	*The Expendable Future: U.S. Politics and the Protection of Biological Diversity.* Durham, NC: Duke University Press, 1991.
Colin Tudge	*Last Animals at the Zoo: How Mass Extinction Can Be Stopped.* Washington: Island Press, 1992.
World Wildlife Fund	*The Official WWF Guide to Endangered Species of North America.* Washington: Beacham Publishing, 1992.

Index

acid rain, 239
African wildlife trade, 184-87, 201
agriculture
  and livestock extinction, 48-55
  and wild plant extinction, 56-62
alligators, 68, 126, 165
American Forest Resource Alliance, 133
American Minor Breeds Conservancy (AMBC), 49-50, 52, 53, 55
American Rails-to-Trails conservancy, 83
Andujar, Claudia, 213
Animals' Voice, 183
Antarctica, whaling in, 174-75, 179
Arabian oryx, 92, 93, 100, 102
Army Corps of Engineers, 155-58, 163-64, 166
artificial insemination, 52-53
Ashby's Law of Requisite Variety, 82
Askins, Renée, 149
Audubon Society, 70, 76, 131, 135
Avery, Dennis, 75-76

Babbitt, Bruce, 125, 147
Balick, Michael, 197
Bean, Michael J., 73
Beck, Ben, 98, 100
Begley, Sharon, 106
Berry, Wendell, 40
biodiversity
  buying rain forest products
    preserves, 193-99
    con, 200-204
  and cultural diversity, 37
  definition of, 81
  deforestation decreases, 33-40
    con, 41-47
  importance of, 25-28, 34-35, 105-109, 115, 179, 226
  protection of, 44-45
  types of, 35
biogeography, 28
biophilia
  is false, 228-33
  importance of, 223-26
biophobia, 224-25, 230-32
biotechnology, 57, 59-61
birds
  Attwater's prairie chicken, 111
  Bachman's warbler, 18, 19
  barn owls, 101

brown pelicans, 68, 76
California condors, 76, 93, 99, 107-109, 114
California gnatcatcher, 70, 84, 87
depend on wetlands, 163
dusky seaside sparrow, 70-71
eagles
  and DDT ban, 68, 75-76
  spending on, 69, 114
  as symbolic, 71, 111, 126
extinction of, 18-19, 38, 67-68, 76, 111
ivory-billed woodpeckers, 18
Kirtland's warbler, 111, 115, 116
peregrine falcons, 68, 69, 76, 114, 115, 126
Puerto Rican parrot, 111, 114, 116
red-cockaded woodpecker, 113-15, 116, 127-28
reintroduction of, 92-93
spending on endangered, 107, 109, 114
spotted owl
  habitat of, 22, 82, 127, 133
  protection of, 21-22, 71
    costs loggers' jobs, 71, 130-35
    con, 136-42
  spending on, 71, 114, 116
Bixby, Don, 49
Body Shop, The, 201, 219
Bohlen, Curtis, 184
Brazil
  deforestation in, 35
  indigenous cultures in, 209, 211-13
Broucek, John, 196, 199
Burke, Kathleen, 51
Burwell, David, 83

California Agricultural Lands Project (CALP), 61
California Natural Communities Conservation Plan, 87
Carlson, Shawn, 41
Carothers, Andre, 183
Carson, Hampton, 116
Center for Plant Conservation, 21
Chase, Alston, 89, 143
Chinantec Indians, 218
Chopra, Sudir K., 178
CITES, 183, 185, 192
Clay, Jason, 195-97, 199, 209, 211
Clean Air Act, 67, 79

Clean Water Act, 155, 163
Clinton, Bill, 82, 86, 140-41
Coastal Sage Scrub ecosystem, 84
Cockburn, Alexander, 136
COICA, 210
Colinvaux, Paul, 28
Communal Areas Management
  Program for Indigenous Resources
  (Campfire), 190-91
conservation, 29-30
  fisheries, 40
  forests, 132, 141, 142
  genetic, 49-50, 53-55, 57
  New Forestry, 39, 135
  protected areas, 36-37, 44
Conservation International (CI), 198
Convention on International Trade in
  Endangered Species (CITES), 183,
  185, 192
corn, 59-60
Corry, Stephen, 200
Cotacachi-Cayapas Ecological
  Reserve, 198
Cultural Survival Enterprises, 194-97,
  201-204, 209
Cushman, John H., Jr., 82

D'Amato, Anthony, 178
Davidson, Art, 208
DDT, ban on, 68, 75-76
De Alessi, Michael, 180
Defenders of Wildlife, 70, 84, 108,
  122, 146
deforestation
  decreases biodiversity, 33-40
    con, 41-47
  and extinction of species, 19, 22,
    34-35
  threatens indigenous people, 209,
    211, 212
  see also rain forests
Delhi Sands fly, 88
desert tortoise, 111, 114, 127
Diamond, Jared, 28, 30, 93, 227
DiSilvestro, Roger L., 104
Dolan, Maura, 22, 38, 100
Douglas-Hamilton, Iain, 184
Drew, Lisa, 48
Durning, Alan Thein, 214
Dutch Belted cattle, 49

Earth Island Institute, 237
Earth Summit, 83, 179
ecosystems
  definition of, 89-90
  degradation of, 35-36, 81
  preservation of
    will save endangered species,
      80-84, 108, 115-16
    will violate property rights, 85-90
  sustainable commercial use of,
    39-40, 121
Egan, Timothy, 136
Ehrlich, Paul, 25, 46, 90
elephants
  harm Africans, 189
  and ivory trade, 184-87, 201
  killing of, 184
  population of, 191-92
  qualities of, 184
  value to ecosystem, 106-107, 184-85
elephant seal, 106
elk, 144, 145
Ellen, William B., 155-59
endangered species
  all should be preserved, 104-109
  humans as, 234-40
  list of, 67-68, 71, 185
    errors in, 67-70, 124
  preserving ecosystems will save,
    80-84
    con, 85-90
  and private property owners
    protection harms, 86-90, 120-24
    regulation is necessary, 125-29
  protection of
    compensation for, 78-79, 122-24,
      146, 150
    is wrong, 126-29
  some cannot be saved, 110-16
  and wetlands, 37, 163, 165
  wildlife reintroduction programs
    will harm, 97-103
    will help, 91-96
  in zoos, 95, 99-102
Endangered Species Act
  costs of, 67, 69, 71, 77, 111, 113-14
  funding for, 77, 86, 111, 113
  is effective, 73-79, 82, 111, 126
  is failure, 66-72, 111, 115-16
  operation of, 67, 71, 86, 111-12, 121
  and private sector, 78-79
  violates private property rights,
    71-72, 86-90, 120-24
  and wolves, 145, 146
Endangered Species Coalition, 70
Environmental Defense Fund, 76
Environmental Protection Agency
  (EPA)
  bans pesticides, 126
  on wetlands, 155, 158, 163-64
Eulenson, Peggy, 196
Everglades, 82
evolution, 223, 228, 235, 245-46
extinction
  avoidance of, 60

benefits of, 27-29, 242
causes of
  deforestation, 19, 22, 34-35
  environmental shocks, 84
  global warming, 83, 84
  habitat destruction, 19, 21, 26, 81, 107
  human, 18-20, 23, 27, 81, 84, 99, 237, 243-44
  introducing new species, 19, 20
  pollution, 20, 84, 126, 237
and evolution, 242-46
forecasts for, 23, 25-26, 29
of indigenous peoples, 37, 209, 211-12
is a serious problem, 17-23
  con, 24-32
rates of, 19, 34, 237, 243
types of, 21-23

Fairness to Land Owners Committee, 155, 164
farming, 49, 52
Fay, John, 89, 112, 113
ferrets, black-footed, 76, 93, 99, 111, 114, 116
Ferris, Robert M., 151
fish
  Chinook salmon, 82, 114
  conservation, 40
  depend on wetlands, 163, 165
  extinction of, 19-20, 74, 77
  reintroduction of, 98
Fitzgerald, Randy, 130
Florida panther, 111, 114, 165
food chain, 38-39
food supply
  genetic diversity needed for, 57, 105-106
Forest Conservation Council, 131, 132
forests
  and acid rain, 239
  boreal, 116
  and clear cutting, 135, 138
  conservation of, 132, 141, 142
  logging in, 131, 137
  New Forestry, 39, 135
  old-growth, 82, 131-33, 142
  see also deforestation; rain forests
Forkan, Patricia A., 171
Frankfurt (Germany) Zoo, 102
Freedom of Information Act, 71-72, 88
frogs, 81, 84

genetic banks, 50, 62
genetic conservation, 49-50, 53-55, 57
genetic diversity, 39, 55

breeding wildlife, 92
  for food supply, 57, 105-106
  landraces, 60
  loss of, 57-58, 60
  Vavilovian centers, 57-59, 61
genetic engineering, 32, 236
genetic erosion, 60, 62
genetic resistance
  livestock, 49-55
  plants, 57-62, 105-106
global warming, 83, 84, 238
golden lion tamarins, 92, 98, 101-102
Goodman, Ellen, 128
Gordon, Robert E., 66, 76, 77, 78, 79
Gore, Al, 19, 56, 237, 239
Gould, Stephen Jay, 46
Greenpeace, 178, 181
Green Revolution, 58, 61
greenways, 82-83
grizzly bear, 107, 111, 114
Grove, Noel, 239

habitats
  destruction of, causes extinction, 19, 21, 26, 81
  fragmentation of, 81, 84
  linking of, 83-84
  protection of saves species, 107, 115
Horton, Tom, 76
Humane Society of the United States, 186
Human Events, 143
humans
  are endangering themselves, 37, 234-40
    con, 241-48
  are innately connected to nature, 222-26
    con, 227-33
  cause extinction, 18-20, 23, 27, 81, 84, 99, 237, 243-44
  dependence on other species, 30, 35, 105, 223-26, 239-40
  habitats of, 225, 228
  living preferences of, 225
  other species adapt to, 26, 29
  as part of life's cycles, 246-48
  phobias of, 224-25, 230-32
  population of
    increases in, 236-37
    threatens plant diversity, 59-61
  promote new species, 27
  starvation of, 189, 192
  transformation of earth by, 223, 235-39, 244-45
  see also indigenous peoples

indicator species, 138, 158

indigenous peoples
  and alcoholism, 210, 212
  are adapting to the modern world, 214-21
  are endangered, 202, 208-13
  connection to nature, 211, 213, 215, 221
  and disease, 211-13
  education for, 210, 217
  knowledge of, 209, 213, 215, 220-21
  land of, 209-10
  maintaining traditional ways, 213, 215, 217-18
  natural resource rights, 216-19
  overuse resources, 216
  political power for, 216-19, 221
  as slaves, 211
  threat to, 202-203
  trade by, 219-20
  using wildlife as a resource, 189-92
insects
  extinction of, 21, 81
  rain forest, 34, 43, 45
International Association for Landscape Ecology, 84
International Council for Bird Preservation, 18, 113
International Fund for Animal Welfare, 182
International Rice Research Institute, 59
International Storage Center for Rice Genes, 57
International Union for Conservation of Nature and Natural Resources, 18, 93-94
International Whaling Commission (IWC), 172-75, 179
Inuit Circumpolar Conference, 218
ivory
  international ban should be lifted, 188-92
  international ban should be maintained, 183-87
  legal trade in, 192

Japan, whaling by, 173-75, 177
Jasper, William F., 154
Jukofsky, Diane, 193

Kangas, Patrick, 44
Kellert, Stephen, 177, 231
Kilpatrick, James J., 78
Kleiman, Devra, 98, 101-102

Lea, Douglass, 80
Lee, Robert W., 31
Lehr, Jay H., 158

Leopold, Aldo, 113, 116
Lewis, Damien, 212
livestock
  ancestry of, 50-51
  benefits of older breeds, 50-52, 53-54
  compensating owners for, 122, 146
  extinction of, 48-55
  modern breeding of, 52
Lovejoy, Thomas, 25, 26, 29, 30
Lugo, Ariel, 27, 28, 46
Luling, Virginia, 212

Mann, Charles C., 112
Marble Mountain Audubon v. Rice, 83
Margulis, Lynn, 243
Mayr, Ernst, 25
McClung, Robert M., 234
Means, Richard L., 239
medicine
  from rain forest plants, 23, 37, 40, 105, 195, 220
  traditional, 37, 40, 105
Meeker-Lowry, Susan, 202
Metrick, Andrew, 69
Miniter, Richard, 26, 165
Minnesota Zoo, 103
Moi, Daniel Arap, 185
mollusks, extinction of, 20-21, 81, 107-108, 113
Monkey Sanctuary, 95-96
Morais, Richard C., 188
mountain lions, 144
Murdock, Deroy, 122
Myers, Norman, 27, 28, 58, 59

National Biological Service (NBS), 88
National Marine Fisheries Service, 67
National Resources Defense Council, 83
National Wetlands Coalition, 164
Native Fish and Wildlife Service, 218
Natural Communities Conservation Planning (NCCP), 84
Nature Conservancy, 92, 113
Navajo Livestock Reduction Program, 54
New England Natural Bakers, 196, 199
Noah's Park, 95
Northwest Forestry Association, 132, 134, 139
Norway, whaling by, 173-74, 177, 179-82
Nugkuag, Evaristo, 210-11

orangutans, 37, 94, 98

Palmer, Thomas, 241

Père David's deer, 92, 93
Perry, Rachel, 196
Peters, Charles M., 38, 194
Pierce, Robert, 158
plants
  extinction of, 21, 36, 56-62
  genetic resistance of, 57-62, 105-106
  indigenous peoples' use of, 195, 215, 219-20
  number of species, 105
  wild
    extinction of threatens agriculture, 56-62
    medicine from, 23, 37, 40, 105, 195, 220
    strains of, 57, 115
Plummer, Mark L., 112
Political Economy Research Center (PERC), 122
Pollot, Mark L., 157
pollution, 20, 84, 126, 237-239
potatoes, 59, 61
Potiguara, Eliane, 210
Pozsgai, John, 159-60, 162
prairies, 35, 107, 116
Przewalski's horse, 51, 92, 93, 99
Pygmies, 212

Rain, Salvador, 215
Rainforest Alliance, 193
Rainforest Crunch, 195, 202
rain forests
  biodiversity in, 42-43
  colonization of, 202
  conservation of, 46-47
  destruction of, 35-36, 46
  and extinction, 19, 21, 25-26, 28, 34
  farming of, 194, 196
  insects in, 34, 43, 45
  land management in, 46-47
  logging in, 194, 196, 204
  medicine from, 37, 40, 105, 195
  operation of, 42
  products from
    buying preserves biodiversity, 193-99
    claims are false, 200-204
  reforestation of, 44, 196
  temperate, 35, 37, 107
Rare Breeds Survival Trust, 49, 52, 53
Raven, Peter, 30, 36
Regan, Tom, 180-82
rhinoceros, 92, 185, 186, 192, 201
rice, 57, 58, 59
Richardson, Valerie, 145
Royal Society for the Protection of Birds, 92
Rubin, Melinda, 196-97

Ryan, John C., 33, 58

Sagan, Dorion, 243
Saxenian, Michael, 198-99
Schindler, Paul R., 185
seals, 106, 181
sea turtles, 74, 76
Serengeti National Park, 102
Shanley, Patricia, 195, 198
Shelby, Richard, 72, 77-79
Sierra Club Legal Defense Fund, 131-32, 145, 237
Simon, Julian L., 24
Skåre, Mari, 176
Smith, Robert J., 120
snails, 20-21
snakes, 224-25, 230-32
Snodgrass, Randall, 76
Soil and Conservation Service, 155-57
Solomon Islands, 18-19
species
  adaptation to humans, 26, 29
  cataloging of, 18, 26, 31, 111
  definition of, 25, 70
  determining extinction of, 18, 28-29, 31, 45, 81, 242
  human promotion of, 27
  interdependency of, 38, 105-107, 240
  island area formula, 26, 42-45
  number of, 26, 34, 81
  see also endangered species
squirrels, 30, 32
Stahl, Andy, 132
Stephens' kangaroo rat, 121-23
Sugg, Ike C., 85
Sunquist, Fiona, 97
sustainable use
  of ecosystems, 39-40, 121
  and rain forest products, 193-99
  and whaling, 173, 177-82

Tauzin, Wilbert J., 72, 74, 75, 78, 79
taxol, 23, 105
technology, effect of, 223, 235-37
Thompson, Douglas A., 161
Tropical Botanicals, 194, 197
Tudge, Colin, 91
Tyunga people, 189-90

United Nations
  Conference on Environment and Development, 83, 179
  endangered species list, 69
  Food and Agriculture Organization, 49
  International Board for Plant Genetic Resources, 60
  Tropical Forestry Action Plan, 47

United States
  Agency for International
    Development, 191
  Department of Agriculture, 50
  diversity center in, 61
  Fish and Wildlife Service (FWS),
    67-69, 74, 77, 108
    Office of Migratory Bird
      Management, 113
    policing private property, 121-22
    saving endangered species, 112-13
    and spotted owls, 131, 132-33,
      137, 138
    and wetlands, 163
    wolf reintroduction program, 145
  government spending for
    endangered species, 107, 113-14
  National Zoo, 98
  should accept commercial whaling,
    176-82
    con, 171-75

van Roosmalen, Marc, 95, 96
Vavilov, Nikolai Ivanovich, 57-59
Vavilovian centers, 57-59, 61

Walker, Donald, Jr., 131, 132, 135
Ward, Barbara, 235
Wearne, Phillip, 219
Weitzman, Martin, 69
Wemmer, Chris, 98, 101, 102
western yew, 22-23
wetlands
  compensation for, 79, 164, 166
  definition of, 157-58, 162
  destruction of
    effects on environment, 163
    legal penalties for, 155, 159
    reasons for, 163, 165
  link to other habitats, 83
  locations of, 162-63
  neglected, 37
  preservation, 116
  regulations are unfair to property
    owners, 154-60
    con, 161-67
whales, 172-73, 178-79
  blue, 172, 178
  bowhead, 180, 215
  gray, 68, 76
  minke, 173-75, 177-81
  U.S. should condemn commercial
    whaling, 171-75
    con, 176-82
whooping crane, 111, 114-16, 165
Wildavsky, Aaron, 24
wildlife

corridors, 82-83
genetic traits in, 92, 94-95, 100
and human starvation, 189, 192
learned behavior of, 94-95, 101
overprotection of, 144
as pets, 229
private ownership of, 189-92
reintroduction plans for
  costs of, 102, 115, 147
  as harmful, 97-103, 143-48
  as helpful, 91-96
safaris, 190
trade in, 95, 185-87, 201, 229
Wilson, Edward O., 31, 81
  on extinction is a serious problem,
    17-23, 25, 90
  on humans as innately connected to
    nature, 222-26
  on humans endangering selves, 237
  on importance of wild species, 105,
    226
  on number of species lost, 26-27,
    34, 45
Winckler, Suzanne, 110
Winner, Langdon, 173
Wolf Reward Program, 122
wolves
  increase in numbers, 145-46
  kill livestock, 145-47
    con, 150, 151
  in Minnesota, 145-46
  in North Carolina, 76, 147-48
  reintroduction threatens ranchers'
    livelihoods, 143-48
    con, 149-53
  spending on, 107, 114
  Tasmanian, 18, 93
  in Yellowstone National Park, 144-47
woolly monkeys, 94, 95-96
World Conservation Union, 98, 102
World Wildlife Fund, 55, 70
Worldwatch Institute, 239

Xapuri Agroextractive Cooperative,
  195

Yanomami, Kopenawa Davi, 212-13
Yanomami people, 211-13
Yellowstone National Park, 144-47,
  151-53
yew trees, 22, 105
Yocom, Thomas G., 161

Zimbabwe Trust, 190
zoos
  endangered species in, 95, 99-102